THE
SUMMATION
OF
SERIES

Harold T. Davis

DOVER PUBLICATIONS, INC.
Mineola, New York

Bibliographical Note

This Dover edition, first published in 2015, is an unabridged republication of the work originally published by The Principia Press of Trinity University, San Antonio, Texas, in 1962.

Library of Congress Cataloging-in-Publication Data

Davis, Harold T. (Harold Thayer), 1892–1974, author.
  The summation of series / Harold T. Davis. — Dover edition.
    pages cm — (Dover books on mathematics)
    Reprint of: San Antonio, Texas : The Principia Press of Trinity University, 1962.
    Includes index.
    Summary: "Valuable as both a text and a reference, this concise monograph was written by a prominent mathematician and educator whose interests encompassed the history of mathematics, statistics, modeling in economics, mathematical physics, and other disciplines. The book is suitable for students, researchers, and applied mathematicians in many areas of mathematics, computer science, and engineering"— Provided by publisher.
    ISBN-13: 978-0-486-78968-2 (paperback)
    ISBN-10: 0-486-78968-3 (paperback)
    1. Series. I. Title.

QA295.D33 2015
515'.243—dc23

2014029950

Manufactured in the United States by LSC Communications
78968302   2017
www.doverpublications.com

DEDICATION

To Richard Gunthorp, without whose sustained interest and help this work would never have been produced, the author affectionately dedicates *The Summation of Series*.

TABLE OF CONTENTS

CHAPTER 4. SUMMATION BY TABLES

CHAPTER 5. INFINITE SUMS

TABLE OF FINITE SUMS

INDEXES

THE
SUMMATION
OF
SERIES

PREFACE

The purpose of this small volume is to advance the reader's understanding of the problem of the summation of series, with special emphasis upon the case of finite limits. In most texts scant attention is paid to this subject, since the emphasis is necessarily upon the techniques of integration. The student thus acquires so little knowledge of the methods of summation that he usually has difficulty in finding the sums of even the simplest functions. Beyond the elementary cases of arithmetic and geometric progressions, his knowledge is of a vague and unsatisfactory kind. It is to remedy this deficiency that the present book is dedicated.

The device used here is to exhibit differences and summations as the analogues of derivatives and integrals in the infinitesimal calculus. Summation formulas are derived as the inverses of differences in the same way that integration formulas are obtained as the inverses of derivatives. The writer thus introduces as his central exhibit a *Table of Finite Sums*, which began with a smaller collection of 41 formulas assembled by Dr. H. E. H. Greenleaf a number of years ago. This collection was later extended by Edwin Greenstein, to which some additions and emendations have been made by the writer.

CHAPTER 1

THE CALCULUS OF FINITE DIFFERENCES

1. Finite Differences. The demands not only of statistics, but of physical science as well, have made desirable some knowledge of what is called the **calculus of finite differences** to distinguish it from the infinitesimal calculus. The calculus of finite differences can be divided into two parts, one concerned with differences, called the **difference** calculus, and the other concerned with summations, called the **summation calculus.** These resemble in many details the differential and integral calculus.

Since the elements of the calculus of finite differences are readily understood by any one who has mastered the principles of the infinitesimal calculus, it will be possible to develop readily many of the important features of this very useful discipline.

By the **difference** of a function $f(x)$, designated by $\Delta f(x)$, or simply Δf, we mean the expression

$$\Delta f(x) = f(x + d) - f(x),$$

where d is the difference interval.

That the differential calculus is in a sense the limiting form of the difference calculus is at once observed from the following limit:

$$D_x f(x) = \lim_{d \to 0} \frac{\Delta f(x)}{d}.$$

In most of the discussion which follows it will simplify the analysis to assume that the difference interval is 1, that is to say, we shall concern ourselves with the special difference defined by

$$\Delta f(x) = f(x + 1) - f(x).$$

Examples of differences are given by the following:

$$\Delta x^2 = (x + 1)^2 - x^2 = 2x + 1,$$

$$\Delta \left(\frac{1}{x} \right) = \frac{1}{x + 1} - \frac{1}{x} = - \frac{1}{x(x + 1)},$$

$$\Delta \sin x = \sin(x + 1) - \sin x = 2 \sin \tfrac{1}{2} \cos (x + \tfrac{1}{2}).$$

The difference calculus can now be developed exactly as the differential calculus was developed by obtaining a set of fundamental formulas. Some of these are obtained by inspection. Thus, denoting by u, v, and w given functions of x, we obtain at once the following:

$$\Delta c = 0, \text{ where } c \text{ is a constant,}$$

$$\Delta(cu) = c\Delta u,$$

$$\Delta(u + v + w) = \Delta u + \Delta v + \Delta w,$$

$$\Delta x = 1.$$

2. The Factorial Symbols. If we form the differences of x^2 and x^3 we obtain $2x + 1$ and $3x^2 + 3x + 1$ respectively, that is, $\Delta x^2 = 2x + 1$ and $\Delta x^3 = 3x^2 + 3x + 1$. These formulas do not have the same simplicity as the corresponding formulas in the differential calculus where $\dfrac{d}{dx} x^2 = 2x$ and $\dfrac{d}{dx} x^3 = 3x^2$. But this same simplicity can be restored to the symbols of the difference calculus if x^n is replaced by what is called the **nth factorial** x, represented by the symbol $x^{(n)}$ and defined as follows:

$$x^{(n)} = x(x - 1)(x - 2) \cdots (x - n + 1), \quad n > 0.$$

Thus we have,

$$\Delta x^{(2)} = \Delta x(x - 1) = (x + 1)x - x(x - 1) = 2x,$$

$$\Delta x^{(3)} = \Delta x(x - 1)(x - 2),$$

$$= (x + 1)x(x - 1) - x(x-1)(x - 2),$$

$$= 3x^{(2)},$$

and in general,

$$\Delta x^{(n)} = [(x + 1)x(x - 1) \cdots (x - n + 2)]$$

$$- [x(x - 1)(x - 2) \cdots (x - n + 1)],$$

$$= [(x + 1) - (x - n + 1)]$$

$$x(x - 1)(x - 2) \cdots (x - n + 2),$$

$$= nx^{(n-1)}.$$

If the factorial index is a negative number, then the **nth reciprocal factorial** x, represented by the symbol $x^{(-n)}$, is defined as follows:

$$x^{(-n)} = \frac{1}{(x + 1)(x + 2) \cdots (x + n)}, \quad n > 0.$$

For $n = 2$ and 3, we get

$$\Delta x^{(-2)} = \Delta \frac{1}{(x+1)(x+2)},$$

$$= \frac{1}{(x+2)(x+3)} - \frac{1}{(x+1)(x+2)},$$

$$= \frac{-2}{(x+1)(x+2)(x+3)} = -2x^{(-3)}.$$

$$\Delta x^{(-3)} = \Delta \frac{1}{(x+1)(x+2)(x+3)},$$

$$= \frac{1}{(x+2)(x+3)(x+4)} - \frac{1}{(x+1)(x+2)(x+3)},$$

$$= \frac{-3}{(x+1)(x+2)(x+3)(x+4)} = -3x^{(-4)};$$

and in general,

$$\Delta x^{(-n)} = \Delta \frac{1}{(x+1)(x+2)\cdots(x+n)},$$

$$= \frac{1}{(x+2)(x+3)\cdots(x+n+1)}$$

$$- \frac{1}{(x+1)(x+2)\cdots(x+n)},$$

$$= \frac{(x+1)-(x+n+1)}{(x+1)(x+2)\cdots(x+n+1)} = -nx^{(-n-1)}.$$

Example 1. Find the difference of

$$u = 2x^{(2)} + 3x^{(4)} - 5x^{(-3)}.$$

Solution: $\Delta u = 2 \Delta x^{(2)} + 3 \Delta x^{(4)} - 5 \Delta x^{(-3)},$

$$= 4x + 12x^{(3)} + 15x^{(-4)}.$$

Example 2. Show that $x^4 = x + 7x^{(2)} + 6x^{(3)} + x^{(4)}$ and from this identity compute the difference of x^4.

Solution: Since both sides of the given equation are polynomials of fourth degree they can be shown to be identical if they coincide for five different values of x. For $x = 0$ and 1 the agreement is obvious. For $x = 2$, we get $16 = 2 + 7 \cdot 2 = 16$; for $x = 3$, $81 = 3 + 7 \cdot 6 + 6 \cdot 6 = 81$; and for $x = 4$, $256 = 4 + 7 \cdot 4 \cdot 3 + 6 \cdot 4 \cdot 3 \cdot 2 + 4 \cdot 3 \cdot 2 \cdot 1 = 4 + 84 + 144 + 24 = 256$. The identity is thus established.

The difference of x^4 is thus found to be

$$\Delta x^4 = \Delta x + 7\Delta x^{(2)} + 6\Delta x^{(3)} + \Delta x^{(4)},$$

$$= 1 + 14x + 18x^{(2)} + 4x^{(3)}.$$

PROBLEMS.

Find the differences of the following functions:

1. x^2. Ans. $2x + 1$.
2. $\dfrac{1}{x^2}$.
3. $\dfrac{x+1}{x+2}$. Ans. $\dfrac{1}{(x+2)\,(x+3)}$.

4. $x^{(2)} - x^{(-2)}$.
5. $x + \dfrac{1}{x}$. Ans. $\dfrac{x^2 + x - 1}{x(x+1)}$.

6. $\dfrac{1}{x(x+1)}$.
7. $\dfrac{1}{x^2 - 1}$. Ans. $\dfrac{-(1+2x)}{(x-1)\,x\,(x+1)\,(x+2)}$.

8. Express x^3 as the sum of factorials. Ans. $x^3 = x^{(3)} + 3x^{(2)} + x$.

9. Use the result of Problem 8 to evaluate Δx^3. Ans. $3x^{(2)} + 6x + 1$.

10. Show that $\Delta \dfrac{x^{(2)}}{x+1} = \dfrac{x(x+3)}{(x+1)\,(x+2)}$.

3. Table of Differences and Its Application. For convenience of application the following table of differences has been formed. This table is analogous to the one given in most works on the differential calculus.

TABLE OF DIFFERENCES.

(1) $\Delta c = 0$, where c is a constant.

(2) $\Delta(cu) = c\Delta u$.

(3) $\Delta x = 1$.

(4) $\Delta(u + v + w) = \Delta u + \Delta v + \Delta w$.

(5) $\Delta u_x v_x = v_{x+1}\Delta u_x + u_x \Delta v_x = u_{x+1}\Delta v_x + v_x \Delta u_x$.

(6) $\Delta\left(\dfrac{u_x}{v_x}\right) = \dfrac{v_x \Delta u_x - u_x \Delta v_x}{v_x v_{x+1}}$.

(7) $\Delta x^{(n)} = nx^{(n-1)}$.

(8) $\Delta x^{(-n)} = -nx^{(-n-1)}$.

(9) $\Delta 2^x = 2^x$.

(10) $\Delta a^x = (a-1)a^x$.

(11) $\Delta \sin(ax + b) = 2 \sin \tfrac{1}{2}a \cos(ax + b + \tfrac{1}{2}a)$.

(12) $\Delta \cos(ax + b) = -2 \sin \tfrac{1}{2}a \sin(ax + b + \tfrac{1}{2}a)$.

(13) $\Delta \tan(ax + b) = \sin a \sec(ax + b) \sec(ax + a + b)$.

(14) $\Delta \cot(ax + b) = -\sin a \csc(ax + b) \csc(ax + a + b)$.

(15) $\Delta\csc(ax+b) = \dfrac{-2\sin\frac{1}{2}a\,\cos(ax+b+\frac{1}{2}a)}{\sin(ax+b)\sin(ax+a+b)}.$

(16) $\Delta\sec(ax+b) = \dfrac{2\sin\frac{1}{2}a\,\sin(ax+b+\frac{1}{2}a)}{\cos(ax+b)\cos(ax+a+b)}.$

Formulas (1) through (4) were discussed in Section 1 and formulas (7) and (8) were derived in Section 2. The others are similarly obtained as shown below.

The difference of the product of two functions as given in formula (5) is derived as follows:

$$\Delta u_x v_x = u_{x+1}v_{x+1} - u_x v_x,$$
$$= u_{x+1}v_{x+1} - u_x v_{x+1} + u_x v_{x+1} - u_x v_x,$$
$$= v_{x+1}(u_{x+1} - u_x) + u_x(v_{x+1} - v_x),$$
$$= v_{x+1}\Delta u_x + u_x\Delta v_x.$$

Since u_x and v_x can be interchanged in this formula, it is clear that we can also write the difference as follows:

$$\Delta u_x v_x = u_{x+1}\Delta v_x + v_x\Delta u_x.$$

The difference of the quotient of two functions is similarly obtained. Thus to establish formula (5) we write

$$\Delta\left(\frac{u_x}{v_x}\right) = \frac{u_{x+1}}{v_{x+1}} - \frac{u_x}{v_x} = \frac{u_{x+1}v_x - v_{x+1}u_x}{v_x v_{x+1}},$$
$$= \frac{u_{x+1}v_x - u_x v_x + u_x v_x - v_{x+1}u_x}{v_x v_{x+1}},$$
$$= \frac{v_x(u_{x+1} - u_x) - u_x(v_{x+1} - v_x)}{v_x v_{x+1}},$$
$$= \frac{v_x\Delta u_x - u_x\Delta v_x}{v_x v_{x+1}}.$$

The function 2^x plays the same role in the difference calculus that e^x does in the differential calculus, for we see from formula (9) that 2^x is unchanged by differencing just as e^x is unchanged by taking its derivative. Thus we have,

$$\Delta 2^x = 2^{x+1} - 2^x = (2-1)\,2^x = 2^x.$$

This is a special case of formula (10) which is readily established as follows:

$$\Delta a^x = a^{x+1} - a^x = (a-1)\,a^x.$$

Formula (11) is derived from the following analysis:

$$\Delta\sin (ax + b) = \sin (ax + a + b) - \sin (ax + b).$$

Referring to tables of trigonometric formulas, we see that the difference of two sines can be written,

$$\sin \theta - \sin \phi = 2 \sin \tfrac{1}{2}(\theta - \phi) \cos \tfrac{1}{2}(\theta + \phi).$$

Applying this formula to the difference $\Delta\sin (ax + b)$, we get

$$\Delta\sin (ax + b) = 2 \sin \tfrac{1}{2}a \cos (ax + b + \tfrac{1}{2}a).$$

Formula (12) is at once derived from this result if we observe that $\sin (ax + b + \tfrac{1}{2}\pi) = \cos (ax + b)$. If, therefore, in formula (11) b is replaced by $b + \tfrac{1}{2}\pi$, formula (12) is obtained.

The derivation of formula (13) comes from an application of (6), since we can write

$$\Delta\tan (ax + b) = \Delta \frac{\sin (ax + b)}{\cos (ax + b)},$$

$$= \frac{\cos (ax + b) [2 \sin \tfrac{1}{2}a \cos (ax + b + \tfrac{1}{2}a)]}{\cos (ax + b) \cos (ax + a + b)}$$

$$+ \frac{\sin (ax + b) [2 \sin \tfrac{1}{2}a \sin (ax + b + \tfrac{1}{2}a)]}{\cos (ax + b) \cos (ax + a + b)},$$

$$= \frac{2 \sin \tfrac{1}{2}a [\cos (ax + b) \cos (ax + b + \tfrac{1}{2}a)}{\cos (ax + b) \cos (ax + a + b)}$$

$$+ \frac{\sin (ax + b) \sin (ax + b + \tfrac{1}{2}a)]}{\cos (ax + b) \cos (ax + a + b)},$$

$$= \frac{2 \sin \tfrac{1}{2}a \cos \tfrac{1}{2}a}{\cos (ax + b) \cos (ax + a + b)},$$

$$= \sin a \sec (ax + b) \sec (ax + a + b).$$

Formula (14) is obtained from this result by replacing b by $b + \tfrac{1}{2}\pi$ and observing that $\cot (ax + b) = -\tan (ax + b + \tfrac{1}{2}\pi)$.

To obtain formula (15) we write

$$\csc (ax + b) = \frac{1}{\sin (ax + b)},$$

and apply (6). Formula (16) then follows by replacing b by $b + \tfrac{1}{2}\pi$ and observing that $\sec (ax + b) = \csc (ax + b + \tfrac{1}{2}\pi)$.

The following examples will illustrate the application of these formulas:

Example 1. Compute the difference of the function

$$u = \frac{x(x-1)(x-2)}{(x+1)(x+2)}.$$

Solution: Since we can write the function in the form $u = x^{(3)} x^{(-2)}$, we apply formula (5) and thus obtain,

$$\Delta u = \Delta x^{(3)} x^{(-2)} = (x+1)^{(-2)} \cdot 3x^{(2)} - x^{(3)} \cdot 2x^{(-3)},$$

$$= \frac{3x(x-1)}{(x+2)(x+3)} - \frac{2x(x-1)(x-2)}{(x+1)(x+2)(x+3)},$$

$$= \frac{3x(x+1)(x-1) - 2x(x-1)(x-2)}{(x+1)(x+2)(x+3)},$$

$$= (x+7) x^{(2)} x^{(-3)}.$$

Example 2. Show that $\Delta u_x^2 = (u_{x+1} + u_x) \Delta u_x$.

Solution: Writing u_x^2 as the product $u_x u_x$ we apply formula (5) as follows:

$$\Delta u_x^2 = \Delta u_x u_x = u_{x+1} \Delta u_x + u_x \Delta u_x = (u_{x+1} + u_x) \Delta u_x.$$

Example 3. Evaluate $\Delta \sin^2 4x$.

Solution: Making use of the result of Example 2, and then applying formula (11), we have

$$\Delta \sin^2 4x = [\sin(4x+4) + \sin 4x][2 \sin 2 \cos(4x+2)],$$

$$= [2 \sin(4x+2) \cos 2][2 \sin 2 \cos(4x+2)],$$

$$= (2 \sin 2 \cos 2)[2 \sin(4x+2) \cos(4x+2)],$$

$$= \sin 4 \sin(8x+4).$$

The same result could be obtained also by writing $\sin^2 4x = \frac{1}{2} - \frac{1}{2} \cos 8x$ and applying formula (12). We thus get

$$\Delta \sin^2 4x = \Delta(\tfrac{1}{2} - \tfrac{1}{2} \cos 8x) = \sin 4 \sin(8x+4).$$

PROBLEMS.

Compute the differences of the following functions:

1. $2x^{(-2)} + 3x^{(2)}$.   *Ans.* $-4x^{(-3)} + 6x$.   2. $\frac{1}{5} x^{(5)} + \frac{1}{x+1} - \frac{1}{3} x^{(-3)}$.

3. $x \cos \pi x$.   *Ans.* $-(1 + 2x) \cos \pi x$.   4. $x \sin \pi x$.

5. $x^{(2)} \sin \pi x$.   *Ans.* $-2x^2 \sin \pi x$.   6. $x^{(2)} \cos \pi x$.

7. $2^x x^{(2)}$.   *Ans.* $2^x [x^{(2)} + 4x]$.   8. $3^x x^{(3)}$.

9. $\sin(\pi x + 5)$.   *Ans.* $-2 \sin(\pi x + 5)$.   10. $\sin(2\pi x + 5)$.

11. $4^x + 4^{-x}$.   *Ans.* $3(4^x - 4^{-x-1})$.   12. $2^x - 2^{-x}$.

13. $2^x \cos \pi x$.   *Ans.* $-3 \cdot 2^x \cos \pi x$.   14. $2^x \sin \pi x$.

15. $\dfrac{1}{\sin \pi x}$.   *Ans.* $-2 \csc \pi x$.   16. $\sec \pi x$.

17. $\dfrac{\cos \pi x}{\sin \pi x}$.   *Ans.* 0.   18. $\sin x \cos x$.

19. $\dfrac{\cos \frac{1}{2}\pi x}{\sin \frac{1}{2}\pi x}$.   *Ans.* $-2 \csc \pi x$.   20. $\dfrac{(x-1)(x-2)}{(x+2)(x+3)}$.

21. $\cos^2 4x$.   *Ans.* $-\sin 4 \sin (8x + 4)$.   22. $\tan^2 2x$.

23. Show that $\Delta (\sin 2\pi x + \sin 4\pi x + \sin 6\pi x + \cdots + \sin 2n\pi x) = 0$.

24. Prove that $\Delta u_x^3 = (u_{x+1}^2 + u_x u_{x+1} + u_x^2) \Delta u_x$.

25. Use the formula in Problem 24 to compute $\Delta \sin^3 \pi x$.

4. *Differences of Higher Order.* The quantity $\Delta^n f(x)$ is called the *n*th difference of $f(x)$, or the difference of order n. It is defined as the difference of $\Delta^{n-1} f(x)$, that is to say, by the equation

$$\Delta^n f(x) = \Delta[\Delta^{n-1} f(x)].$$

Thus we have

$$\Delta^2 f(x) = \Delta[\Delta f(x)] = \Delta[f(x+1) - f(x)]$$
$$= f(x+2) - 2f(x+1) + f(x),$$
$$\Delta^3 f(x) = \Delta[\Delta^2 f(x)] = f(x+3) - 3f(x+2)$$
$$+ 3f(x+1) - f(x),$$

and in general,

$$\Delta^n f(x) = f(x+n) - nf(x+n-1)$$
$$+ \frac{n(n-1)}{2!} f(x+n-2) + \cdots + (-1)^n f(x).$$

It is convenient in connection with differences of higher order to introduce the new symbol: $E f(x)$, defined by the equation

$$E f(x) = (1 + \Delta) f(x) = f(x) + \Delta f(x) = f(x+1).$$

We then have

$$E^2 f(x) = (1 + \Delta)^2 f(x)$$
$$= f(x) + 2\Delta f(x) + \Delta^2 f(x) = f(x+2),$$

and, in general,

$$E^n f(x) = (1 + \Delta)^n f(x) = f(x+n).$$

The foregoing formulas can be extended without difficulty to the difference $\Delta f(x) = f(x + d) - f(x)$. We thus have

$$\Delta^n f(x) = f(x + nd) - nf([x + (n - 1)d]$$

$$+ \frac{n(n - 1)}{2!} f[x + (n - 2)d] + \cdots + (-1)^n f(x), \quad (1)$$

and the companion formula,

$$E^n f(x) = (1 + \Delta)^n f(x) = f(x + nd). \quad (2)$$

5. *The Gregory-Newton Interpolation Formula.* Let us now generalize formula (2) of the preceding section by replacing n by p, where p is assumed to have any positive or negative value. We can then write

$$E^p f(x) = f(x + pd) = (1 + \Delta)^p f(x),$$

$$= [1 + p\,\Delta + \frac{p\,(p - 1)}{2!} \Delta^2 + \cdots] f(x),$$

that is to say,

$$f(x + pd) = f(x) + p\,\Delta f(x) + \frac{p\,(p - 1)}{2!} \Delta^2 f(x)$$

$$+ \frac{p\,(p - 1)\,(p - 2)}{3!} \Delta^3 f(x) + \cdots \quad (1)$$

This famous expression is the so-called **Gregory-Newton inter-polation formula,** first announced by James Gregory (1638-1675) in 1670 and later published by Newton in his *Principia Mathematica* (1687). This formula assumes the same role in the difference calculus as that assumed by Taylor's theorem in the differential calculus.

Introducing factorial polynomials, the analogous character of the two formulas is immediately observed, for we have

$$f(x + pd) = f(x) + p\,\Delta f(x) + \frac{p^{(2)}}{2!} \Delta^2 f(x)$$

$$+ \frac{p^{(3)}}{3!} \Delta^3 f(x) + \cdots, \quad (2)$$

in the case of the Gregory-Newton formula, and

$$f(x + pd) = f(x) + pd\,Df(x) + \frac{p^2 d^2}{2!} D^2 f(x)$$

$$+ \frac{p^3 d^3}{3!} D^3 f(x) + \cdots, \quad (3)$$

where D, D^2, D^3, etc. denote derivatives of first, second, third, etc. orders, in the case of Taylor's series.

If, in (2), we set $d = 1$ and $x = 0$, then we get what is the analogue of the well known Maclaurin's series, namely,

$$f(p) = f(0) + p\Delta f(0) + \frac{p^{(2)}}{2!} \Delta^2 f(0) + \frac{p^{(3)}}{3!} \Delta^3 f(0) + \cdots \quad (4)$$

This is to be compared with Maclaurin's series

$$f(p) = f(0) + p\, Df(0) + \frac{p^2}{2!} D^2 f(0) + \frac{p^3}{3!} D^3 f(0) + \cdots, \quad (5)$$

where, as before D, D^2, D^3, etc. denote derivatives of first, second, third, etc. orders.

If, in particular, we let $f(p) = p^n$ in formula (4), then the expansion is in terms of the quantities,

$$\Delta^r 0^n = \lim_{x \to 0} \Delta^r x^n,$$

which are called the **differences of zero**. Since the differences of zero divided by $r!$ have special usefulness, we give below a short table of $(\Delta^r 0^n)/r!$

Values of $\dfrac{\Delta^r 0^n}{r!}$.

	0^1	0^2	0^3	0^4	0^5	0^6	0^7	0^8	0^9	0^{10}
$\Delta\ /1!$	1	1	1	1	1	1	1	1	1	1
$\Delta^2 /2!$		1	3	7	15	31	63	127	255	511
$\Delta^3 /3!$			1	6	25	90	301	966	3025	9330
$\Delta^4 /4!$				1	10	65	350	1701	7770	34105
$\Delta^5 /5!$					1	15	140	1050	6951	42525
$\Delta^6 /6!$						1	21	266	2646	22827
$\Delta^7 /7!$							1	28	462	5880
$\Delta^8 /8!$								1	36	750
$\Delta^9 /9!$									1	45
$\Delta^{10}/10!$										1

By means of these differences of zero one can readily find the difference of x^n. Thus we can write

$$x^n = \sum_{r=1}^{n} x^{(r)} \frac{\Delta^r 0^n}{r!}. \quad (6)$$

Therefore, taking the difference of both sides of the equation, we get

$$\Delta x^n = \sum_{r=1}^{n} x^{(r-1)} \frac{\Delta^r 0^n}{(r-1)!}.$$

The following examples will illustrate the application of the formulas developed in this section:

Example 1. The following table gives the values to four decimal places of cos x from $x = 0.0$ to $x = 1.6$ at intervals of $d = 0.2$, together with differences to third order. Use this table to compute the value of cos 0.5.

x	$\cos x$	Δ	Δ^2	Δ^3
0.0	1.0000			
0.2	0.9801	−0.0199		
		−0.0590	−0.0391	
0.4	*0.9211*		−0.0368	0.0023
		−*0.0958*		0.0040
0.6	0.8253		−*0.0328*	
		−0.1286		*0.0050*
0.8	0.6967		−0.0278	0.0063
		−0.1564		
1.0	0.5403		−0.0215	0.0070
		−0.1779		
1.2	0.3624		−0.0145	0.0077
		−0.1924		
1.4	0.1700		−0.0068	
		−0.1992		
1.6	−0.0292			

Solution: We make use of formula (1). Since the value of cos x is desired at the point $x = 0.5$, we first compute $p = \dfrac{0.5 - 0.4}{(d = 0.2)} = \frac{1}{2}$. The differences to be used are those in italics. We thus obtain

$$\cos 0.5 = 0.9211 + \tfrac{1}{2}(-0.0958) + \frac{\tfrac{1}{2}(\tfrac{1}{2} - 1)}{2!}(-0.0328)$$

$$+ \frac{\tfrac{1}{2}(\tfrac{1}{2} - 1)(\tfrac{1}{2} - 2)}{3!}(0.0050),$$

$$= 0.9211 - 0.0479 + 0.0041 + 0.0003 = 0.8776,$$

which is correct to the last place.

Example 2. Compute the first and second differences of the function

$$y = x^4 + 3x^3 - 2x^2 + 4x.$$

Solution: We first express x^2, x^3 and x^4 in terms of factorials. Thus, making use of formula (6) and the table of the difference of zero, we have

$$x^2 = x + x^{(2)}, \quad x^3 = x + 3x^{(2)} + x^{(3)}, \quad x^4 = x + 7x^{(2)} + 6x^{(3)} + x^{(4)}.$$

Introducing these values into the given function, we can write y as follows:

$y = x^{(4)} + 9x^{(3)} + 14x^{(2)} + 6x$. Taking first and second differences of this function we get

$$\Delta y = 4x^{(3)} + 27x^{(2)} + 28x + 6, \quad \Delta^2 y = 12x^{(2)} + 54x + 28.$$

PROBLEMS.

Compute the second differences of the following functions:

1. $2^x \cos \pi x$.   *Ans.* 9 $2^x \cos \pi x$.   2. $x \cdot 4^x$.

3. $2^x x^{(2)}$.   *Ans.* $2^x [x^{(2)} + 8x + 8]$.   4. $x^3 + 2x^2 + x$.

5. $\sin^2 4x$.   *Ans.* $2 \sin^2 4 \cos (8x + 8)$.   6. $\csc \pi x$.   *Ans.* $4 \csc \pi x$.

7. Expand $y = 3x^4 + 2x^3 - x$ in terms of factorials and compute $\Delta^2 y$.

$$\textit{Ans. } 36x^{(2)} + 120x + 54.$$

8. Given $y = x^5$, show that $\Delta^3 y = 60x^{(2)} + 240x + 150$.

9. Use the table of Example 1 to compute $\cos 0.88^r$. *Ans.* 0.6372.

10. Use the following table of values to compute $\sqrt{2.5}$:

x	$\sqrt{x}$	x	$\sqrt{x}$
1	1.0000	4	2.0000
2	1.4142	5	2.2361
3	1.7321	6	2.4495

$$\textit{Ans. } 1.581.$$

11. Compute the differences of the function $y = x^3$ at $x = 3$ and use these values in series (1) to compute x^3 at $x = 3.5$.

12. Verify the values: $\Delta^3 0^3 = 6$, $\Delta^2 0^4 = 14$.

6. *Summations.* Exactly as in the infinitesimal calculus, the operation inverse to differencing is summation. For example, since we have $\Delta x^{(n)} = n x^{(n-1)}$, it is clear that we shall obtain by the inverse operation the formula

$$\sum x^{(n)} = \frac{x^{(n+1)}}{n+1} + C,$$

where C is an arbitrary constant.

This is readily verified by taking the difference of the right-hand member as follows:

$$\Delta \left\{ \frac{x^{(n+1)}}{n+1} + C \right\} = \Delta \left\{ \frac{x^{(n+1)}}{n+1} \right\} + \Delta C = x^{(n)} + 0.$$

The following table summarizes a few of these useful formulas. They can be verified by taking the differences of both sides of the equations, using for this purpose the *Table of Differences* given in Section 3:

TABLE OF SUMMATIONS.

(1) $\quad \sum a = ax + C.$

(2) $\sum cu = c \sum u.$

(3) $\sum (u + v + w) = \sum u + \sum v + \sum w.$

(4) $\sum x = \frac{1}{2}x^{(2)} + C.$

(5) $\sum x^{(n)} = \frac{x^{(n+1)}}{n + 1} + C, n > 0.$

(6) $\sum x^{(-n)} = \frac{x^{(-n+1)}}{-n + 1} + C, n \neq 1.$

(7) $\sum 2^x = 2^x + C.$

(8) $\sum a^x = \frac{a^x}{a - 1} + C, a \neq 1.$

(9) $\sum \sin(ax + b) = -\frac{\cos(ax + b - \frac{1}{2}a)}{2 \sin \frac{1}{2}a} + C.$

(10) $\sum \cos(ax + b) = \frac{\sin(ax + b - \frac{1}{2}a)}{2 \sin \frac{1}{2}a} + C.$

(11) $\sum a^x \sin kx = a^x \frac{a \sin k(x - 1) - \sin kx}{a^2 - 2a \cos k + 1} + C.$

(12) $\sum a^x \cos kx = a^x \frac{a \cos k(x - 1) - \cos kx}{a^2 - 2a \cos k + 1} + C.$

(13) $\sum \sec(ax + b) \sec(ax + a + b) = \frac{\tan(ax + b)}{\sin a} + C.$

(14) $\sum \csc(ax + b) \csc(ax + a + b)$

$= -\frac{\cot(ax + b)}{\sin a} + C.$

(15) $\sum u_x \Delta v_x = u_x v_x - \sum v_{x+1} \Delta u_x.$

In the *Table of Summation* each indefinite sum has introduced an arbitrary constant C as in integration. The situation here, however, is not exactly comparable with integration since any arbitrary function of unit period can be used in place of C. Thus if $u(x)$ is such a function, then $\Delta u(x) = u(x + 1) - u(x) = 0$. In most applications of this calculus, however, it is sufficient to represent the arbitrary element by C.

The formulas just given find immediate use in the summation of series. The following sum indicates the nature of the application:

$$\sum_{x=m}^{n} \Delta f(x) = \sum_{x=m}^{n} [f(x+1) - f(x)],$$

$$= [f(m+1) - f(m)] + [f(m+2) - f(m+1)]$$
$$+ \cdots + [f(n+1) - f(n)],$$

$$= f(n+1) - f(m).$$

We thus obtain the formula

$$\sum_{x=m}^{n} \Delta f(x) = f(x) \Big|_{m}^{n+1} = f(n+1) - f(m),$$

in which we observe that the upper limit of the summation sign is increased by one unit in the final substitution.

The following examples will illustrate the application of the summation formulas:

Example 1. Find the sum of the squares of the first n integers.

Solution: We first write x^2 in terms of factorials, $x^2 = x^{(2)} + x$. We then have

$$1^2 + 2^2 + 3^2 + \cdots + n^2 = \sum_{1}^{n} x^2 = \sum_{1}^{n} [x^{(2)} + x]$$

$$= \left[\frac{1}{3} x^{(3)} + \frac{1}{2} x^{(2)}\right]_{1}^{n+1} = \frac{n(n+1)(2n+1)}{6}.$$

Example 2. Prove that the sum of the cubes of the first n integers is a perfect square.

Solution: Writing $x^3 = x^{(3)} + 3x^{(2)} + x$, we have

$$\sum_{1}^{n} x^3 = \sum_{1}^{n} [x^{(3)} + 3x^{(2)} + x] = \left[\frac{1}{4}x^{(4)} + x^{(3)} + \frac{1}{2}x^{(2)}\right]_{1}^{n+1}$$

$$= [\tfrac{1}{2}n(n+1)]^2.$$

Example 3. Prove the following:

$$\frac{1}{1\cdot2} + \frac{1}{2\cdot3} + \frac{1}{3\cdot4} + \cdots + \frac{1}{n(n+1)} = \frac{n}{n+1}.$$

Solution: Introducing the reciprocal factorial $x^{(-2)}$, we get

$$\sum_{0}^{n-1} x^{(-2)} = \sum_{0}^{n-1} \frac{1}{(x+1)(x+2)} = -x^{(-1)}\Big|_{0}^{n} = 1 - \frac{1}{n+1} = \frac{n}{n+1}.$$

CALCULUS OF FINITE DIFFERENCES

Example 4. Establish the following identity:

$$\sin\theta + \sin 2\theta + \sin 3\theta + \cdots + \sin n\theta = \frac{\sin \frac{1}{2}(n+1)\theta \, \sin \frac{1}{2}n\theta}{\sin \frac{1}{2}\theta}.$$

Solution: Since the series is the sum of $\sin x\theta$ from 1 to n, we make use of formula (9) in the *Table of Summations* and thus obtain

$$\sum_{1}^{n} \sin x\theta = \frac{-\cos(x\theta - \frac{1}{2}\theta)}{2\sin \frac{1}{2}\theta}\bigg|_{1}^{n+1} = \frac{-\cos(n\theta + \frac{1}{2}\theta) + \cos \frac{1}{2}\theta}{2\sin \frac{1}{2}\theta},$$

$$= \frac{\sin \frac{1}{2}(n+1)\theta \, \sin \frac{1}{2}n\theta}{\sin \frac{1}{2}\theta}.$$

Example 5. Evaluate the sum: $\sum_{1}^{n} x^{(2)} 2^x$.

Solution: To find this sum we make use of summation by parts, formula (15) in the *Table of Summations*. Thus, in the formula

$$\sum u_x \Delta v_x = u_x v_x - \sum v_{x+1} \Delta u_x,$$

we identify u_x with $x^{(2)}$ and Δv_x with 2^x, from which we have $v_x = 2^x$. Hence, we get

$$\sum x^{(2)} 2^x = x^{(2)} 2^x - \sum 2^{x+1} \cdot 2x.$$

We now apply summation by parts a second time to the sum

$$\sum 2^{x+1} \cdot 2x = 4 \sum x \, 2^x,$$

in which we identify u_x with x *and* Δv_x with 2^x. We thus have

$$\sum 2^{x+1} \cdot 2x = 4 \sum x \, 2^x = 4x \cdot 2^x - 4 \cdot 2^{x+1} + C.$$

Combining these sums, we then obtain

$$\sum_{1}^{n} x^{(2)} 2^x = 2^x[x^{(2)} - 4x + 8]\bigg|_{1}^{n+1} = 2^{n+1}(n^2 - 3n + 4) - 8.$$

PROBLEMS.

Compute the value of the following sums:

1. $\sum (x^2 + x)$. *Ans.* $\frac{1}{3}x^{(3)} + x^{(2)} + C$.

2. $\sum \frac{1}{x(x+1)}$.

3. $\sum (x^3 + 2x^2)$. *Ans.* $\frac{1}{4}x^{(4)} + \frac{5}{3}x^{(3)} + \frac{3}{2}x^{(2)} + C$.

4. $\sum (3^x + 5^x + 7^x)$.

5. $\sum (x+1)(x-1)$. *Ans.* $\frac{1}{3}x^{(3)} + \frac{1}{2}x^{(2)} - x + C$.

6. $\sum (4^x + 4^{-x})$.

7. $\sum (a^x - a^{-x})$. *Ans.* $\dfrac{1}{a-1}(a^x + a^{-x+1}) + C$.   8. $\sum \sin \pi x$.

9. $\sum \sec \frac{1}{2}\pi x \csc \frac{1}{2}\pi x$. *Ans.* $- \tan \frac{1}{2}\pi x + C$.   10. $\sum x \cdot 2^x$.

11. $\sum x \cdot 3^x$. *Ans.* $\frac{1}{4}(2x - 3) 3^x + C$.   12. $\sum (x^{(-2)} + x^{(2)})$.

13. $\sum 2^x \sin \pi x$. *Ans.* $- \dfrac{1}{3} 2^x \sin \pi x + C$.   14. $\sum \dfrac{1}{(x+2)(x+1)}$.

15. $\sum 4^x \cos \pi x$. *Ans.* $- \dfrac{1}{5} 4^x \cos \pi x + C$.

16. $\sum (x + a)(x + b)$. *Ans.* $\dfrac{1}{3} x^{(3)} + \frac{1}{2}(a + b + 1) x^{(2)} + abx + C$.

17. $\sum a^x (\sin \pi x + \cos \pi x)$. *Ans.* $- \dfrac{a^x}{a+1} (\sin \pi x + \cos \pi x)$.

18. $\sum \csc \frac{1}{2}\pi x \csc (\frac{1}{2}\pi x + \frac{1}{2}\pi)$. *Ans.* $- \cot \frac{1}{2}\pi x + C$.

19. $\sum x^4$. *Ans.* $\dfrac{1}{5} x^{(5)} + \dfrac{3}{2} x^{(4)} + \dfrac{7}{3} x^{(3)} + \dfrac{1}{2} x^{(2)} + C$.

20. $\sum \sin^2 x$. *Ans.* $\frac{1}{2}x - \frac{1}{4} \dfrac{\sin (2x - 1)}{\sin 1} + C$.

Prove the following:

21. $\displaystyle\sum_1^n x^{(3)} = \frac{1}{4}n(n - 2)(n^2 - 1)$.

22. $\displaystyle\sum_1^{100} [x^2 - x^{(2)}] = 5{,}050$.

23. $\displaystyle\sum_1^n (6x^2 - 6x + 2) = 2n^3$.

24. $\displaystyle\sum_1^n (ax^2 + bx + c) = \frac{n}{6}\Big[(2n^2 + 3n + 1)a + 3nb + 3b + 6c\Big]$.

25. $\displaystyle\sum_1^{10} x^{(2)} 2^x = 152 \cdot 997$.

26. $\displaystyle\sum_1^{10} 2^x \sin \frac{1}{2}\pi x = 410$.

27. $1 + r + r^2 + r^3 + \cdots + r^n = \dfrac{1 - r^{n+1}}{1 - r}$, $r \neq 1$.

28. $r + 2r^2 + 3r^3 + \cdots n\, r^n = \dfrac{r + (nr - n - 1)r^{n+1}}{(1 - r)^2}$, $r \neq 1$.

29. $\dfrac{1}{1 \cdot 2 \cdot 3} + \dfrac{1}{2 \cdot 3 \cdot 4} + \dfrac{1}{3 \cdot 4 \cdot 5} + \dfrac{1}{4 \cdot 5 \cdot 6} + \cdots = \frac{1}{4}$.

30. $1 \cdot 2 \cdot 3 + 2 \cdot 3 \cdot 4 + 3 \cdot 4 \cdot 5 + \cdots + n(n + 1)\,(n + 2)$

$= \frac{1}{4}n(n + 1)\,(n + 2)\,(n + 3)$.

7. Operational Devices in the Calculus of Finite Differences.
In Sections 4 and 5 certain expressions were introduced which, when developed by the ordinary rules of algebra, were still found to preserve their validity as operators. These operational symbols were E^n and $(1 + \Delta)^n$. We shall now relate these symbols to the derivative symbol D and in this manner derive some remarkable relationships between the infinitesimal calculus and the calculus of finite differences.

To show this operational equivalence, let us observe that the development of the function $f(x + d)$ by Taylor's theorem can be written symbolically as follows:

$$f(x + d) = f(x) + d\, f'(x) + \frac{d^2\, f''(x)}{2\,!} + \frac{d^3\, f^{(3)}(x)}{3\,!} + \cdots ,$$

$$= \left(1 + dD + \frac{d^2 D^2}{2\,!} + \frac{d^3 D^3}{3\,!} + \cdots \right) f(x), \qquad (1)$$

where D^2 denotes the second derivative, D^3 the third derivative, and so on.

But the quantity within the parentheses in (1) is formally the expansion of e^{dD}, and in an operational sense we can write

$$f(x + d) = e^{dD}\, f(x).$$

But since we also have $\Delta f(x) = f(x + d) - f(x)$, we can write

$$\Delta f(x) = (e^{dD} - 1)\, f(x), \qquad (2)$$

from which we get, by equating the operators themselves, the equivalence

$$\Delta = e^{dD} - 1. \qquad (3)$$

Similarly, we have

$$E = 1 + \Delta = e^{dD}, \qquad (4)$$

from which we get

$$D = \frac{1}{d} \log (1 + \Delta) = \frac{1}{d}\left[\Delta - \frac{\Delta^2}{2} + \frac{\Delta^3}{3} - \frac{\Delta^4}{4} + \cdots\right]. \quad (5)$$

This last formula is of importance in computing the derivative of a function which is defined by tabulated values. Thus, in the following table, the values of sin x are given, where x is measured in radians:

x	$\sin x$	Δ	Δ^2	Δ^3
1.0	0.8415			
1.1	0.8912	0.0497	−0.0089	
1.2	0.9320	0.0408	−0.0092	−0.0003
1.3	0.9636	0.0316		

To compute the derivative of sin x at the point $x = 1$, we substitute sin x for $f(x)$ in the formula

$$Df(x) = \frac{df(x)}{dx} = \frac{1}{d}\left[\Delta f(x) - \frac{1}{2}\Delta^2 f(x) + \frac{1}{3}\Delta^3 f(x)\right.$$

$$\left. - \frac{1}{4}\Delta^4 f(x) + \cdots\right], \quad (6)$$

where d, the tabular interval, equals 0.1 and the differences are the values 0.0497, −0.0089, −0.003. Since $D \sin x = \cos x$, we thus obtain

$$\cos 1.0 = 10 \left(0.0497 + \frac{1}{2} \times 0.0089 - \frac{1}{3} \times 0.0003\right) = 0.5405,$$

which differs from the correct value 0.5403 in the last place.

Another useful expansion is obtained from the formal expansion of the reciprocal operator obtained by inverting equation, (3) that is

$$\Delta^{-1} = \frac{1}{e^{dD} - 1}. \quad (7)$$

The function $D/(e^{dD} - 1)$ can be expanded in a Maclaurin's series by means of formula (5), Section 5. This expansion was first made by James Bernoulli (1654-1705) in his *Ars Conjectandi* (*The Art of Probability*), published posthumously in 1713. In this expansion certain numbers are introduced, which we denote by B_n and which are called *Bernoulli's numbers*. Thus, we get

$$\frac{D}{e^{dD} - 1} = \frac{1}{d}\left[1 - \frac{1}{2}dD + \frac{B_1 d^2 D^2}{2!} - \frac{B_2 d^4 D^4}{4!}\right.$$

$$\left. + \frac{B_3 d^6 D^6}{6!} - \frac{B_4 d^8 D^8}{8!} + \cdots\right],$$

where $B_1 = 1/6$, $B_2 = 1/30$, $B_3 = 1/42$, $B_4 = 1/30$, $B_5 = 5/66$, $B_6 = 691/2730$, $\cdots$.

The expansion of (7) by means of this series can now be written as follows:

$$\Delta^{-1} = \frac{1}{d} D^{-1} - \frac{1}{2} + \frac{B_1 dD}{2!} - \frac{B_2 d^3 D^3}{4!} + \frac{B_3 d^5 D^5}{6!} - \frac{B_4 d^7 D^7}{8!} + \cdots .$$

This formula can be interpreted operationally if we set $\Delta^{-1} = \sum$, and $D^{-1} = \int$. Thus we have formally

$$\sum f(x) = \frac{1}{d} \int^x f(x) \, dx - \tfrac{1}{2} f(x) + \frac{B_1}{2} df'(x)$$

$$- \frac{B_2}{4!} d^3 f^{(3)}(x) + \frac{B_3}{6!} d^5 f^{(5)}(x) + \cdots \qquad (8)$$

Example: As an example of the application of (8), let us derive the formula for the sum of the squares of the first n integers, which we have already given in Example 1 of Section 6.

Thus, by means of formula (8), letting $f(x) = x^2$, and $d = 1$, we get

$$\sum x^2 = \frac{1}{3} x^3 - \frac{1}{2} x^2 + \frac{1}{6} x + C = \frac{x(x-1)(2x-1)}{6} + C.$$

The value of the squares of the first n integers follows immediately from this formula, since we have

$$\sum_{x=1}^{n} x^2 = \left[\frac{x(x-1)(2x-1)}{6} + C \right]_1^{n+1} = \frac{n(n+1)(2n+1)}{6}.$$

PROBLEMS.

1. Form the differences Δx^3, $\Delta^2 x^3$, and $\Delta^3 x^3$. Substitute these in the right-hand member of equation (6) and show that the sum reduces to $3x^2$.

2. Compute the value of sin 0.2 by computing the derivative of cos x at $x = 0.2$ from the table given in Example 1 of Section 5.

3. Form a table with differences of the function log x at the points $x = 2.0, 2.1, 2.2, 2.3,$ and 2.4. From the table so constructed compute the derivative of log x at the point $x = 2.1$. *Ans.* 0.477.

4. Form a table with differences of the function sinh u at the points $u = 0.2, 0.3, 0.4, 0.5,$ and 0.6. From the table so constructed evaluate cosh u at $u = 0.2$ by forming the derivative of sinh u at that point.

5. The second derivative of $f(x)$ can be computed from the formula

$$D^2 f(x) = \frac{1}{d^2} \left[\Delta^2 f - \Delta^3 f + \frac{11}{12} \Delta^4 f - \frac{5}{6} \Delta^5 f + \cdots \right].$$

Show that this formula is obtained by squaring both members of formula (5).

6. Use the formula in Problem 5 and the table of values of sin x given in this section to find the second derivative of sin x at $x = 1$.

7. Use formula (8) to prove that: $\displaystyle\sum_{x=1}^{n} x^3 = [\tfrac{1}{2}n(n+1)]^2$.

8. Use formula (8) to compute the sum: $\displaystyle\sum_{x=1}^{n} (1 + x + x^2)$.

9. Substitute $1/x$ for $f(x)$ in formula (8) and show that the right-hand member reduces to the series

$$\log x - \frac{1}{2x} - \frac{1}{12x^2} + \frac{1}{120x^4} + \cdots .$$

10. Although it can be proved by methods beyond this book that the series in Problem 9 does not coverge for any value of x, it can be used, nevertheless, to compute the value of the sum $\displaystyle\sum \frac{1}{x}$ to a high order of approximation if x is sufficiently large. Use the series to compute the value of the sum: $\displaystyle\sum_{x=100}^{199} \frac{1}{x}$.

8. The Euler-Maclaurin Formula for Numerical Integration. By the means of the expansion (8) given in the preceding section it is possible to derive a celebrated formula for numerical integration. This formula was discovered independently by L. Euler (1707-1783) and C. Maclaurin (1698-1746) around 1740 and is thus named after both of them.

The formula can be written conveniently in the following form:

$$\frac{1}{d}\int_a^{a+nd} f(x)\, dx = \sum_{x=0}^{n} f(a + xd) - \frac{1}{2}\left[f(a) + f(a + nd) \right]$$

$$- \frac{d}{12}\left[f'(a + nd) - f'(a) \right] + \frac{d^3}{720}\left[f^{(3)}(a + nd) - f^{(3)}(a) \right]$$

$$- \frac{d^5}{30240}\left[f^{(5)}(a + nd) - f^{(5)}(a) \right] + \cdots$$

$$+ (-1)^m \frac{B_m d^{2m-1}}{(2m)!}\left[f^{(2m-1)}(a + nd) - f^{(2m-1)}(a) \right] + R_m, \quad (1)$$

where B_m is the mth Bernoulli number and R_m can be estimated from the inequality:

$$|R_m| < n\, \frac{B_{m+1}}{(2m + 2)!}\, d^{2m+2} |f^{(2m+2)}(\xi)|,$$

in which ξ is some value between a and $a + nd$.

The expansion which has just been given is obtained immediately from formula (8) of Section 7 by assuming appropriate limits for the sum. Thus we get:

$$\sum_{x=a}^{a+(n-1)d} f(x) = \frac{1}{d} \int_a^{a+nd} f(x) \, dx - \tfrac{1}{2}[f(a + nd) - f(a)]$$

$$+ \frac{B_1}{2!} d \, [f'(a + nd) - f'(a)] + \cdots. \quad (2)$$

If we now replace the Bernoulli numbers by their numerical values and observe that:

$$\sum_{x=a}^{a+(n-1)d} f(x) + \tfrac{1}{2}[f(a + nd) - f(a)]$$

$$= \sum_{x=a}^{a+nd} f(x) - \tfrac{1}{2}[f(a) + f(a + nd)],$$

formula (1) is obtained by expressing the integral in (2) in terms of the other quantities.

It is not possible to obtain the estimate of the remainder R_m by a simple argument and the reader is referred to more advanced books for this derivation.

The Euler-Maclaurin formula is immediately seen to be an extension of the *trapezoidal rule* for integral approximation, found in most books on calculus, since the first three terms of (1) are identical with the terms of that rule. The statement is sometimes made without proof that the approximation by the trapezoidal rule is improved if a term proportional to the difference of the values of the derivative at the ends of the interval of integration is subtracted. We now see that this statement is derived from the inclusion in the approximation by the trapezoidal rule of the fourth and fifth terms of the Euler-Maclaurin formula.

The following examples illustrate the use of the Euler-Maclaurin formula.

Example 1. Use the following table to compute the value of the Fresnel integral: $\int_0^1 \sin x^2 \, dx$.

x	$\sin x^2$	$\cos x^2$	x	$\sin x^2$	$\cos x^2$
0.0	0.00000	1.00000	0.6	0.35227	0.93590
0.1	0.01000	0.99995	0.7	0.47063	0.88233
0.2	0.03999	0.99920	0.8	0.59720	0.80210
0.3	0.08988	0.99595	0.9	0.72429	0.68950
0.4	0.15932	0.98723	1.0	0.84147	0.54030
0.5	0.24740	0.96891	Totals	3.53245	9.80137

Solution: Since $f(x) = \sin x^2$, we first compute $f'(x) = 2x \cos x^2$, $f^{(3)}(x) = -12x \sin x^2 - 8x^3 \cos x^2$. These functions are all zero when $x = 0$, but at $x = 1$ they have values- $f(1) = 0.84147$, $f'(1) = 1.08060$, $f^{(3)}(1) = -14.4200$. Observing that $d = 0.1$, we now introduce these values and the sum of $\sin x^2$, namely 3.53245, into formula (1) and thus obtain:

$$\int_0^1 \sin x^2 dx = 0.1 \left(3.53245 - \tfrac{1}{2}\, 0.84147 - \frac{0.1}{12}\, 1.08060 - \frac{0.001}{720}\, 14.4200\right),$$

$$= 0.1\,(3.53245 - 0.42074 - 0.00901 - 0.00002) = 0.31027.$$

This value is correct to the last place.

Example 2. Find the value of the sum: $S = \dfrac{1}{100} + \dfrac{1}{101} + \dfrac{1}{102} + \cdots + \dfrac{1}{200}$.

Solution: To find the desired sum we introduce the function $f(x) = \dfrac{1}{x}$ into formula (1), letting $d = 1$, $a = 100$ and $a + nd = 200$. Since $f'(x) = -\dfrac{1}{x^2}$ and $f^{(3)}(x) = -\dfrac{6}{x^4}$, we thus obtain:

$$\int_{100}^{200} \frac{dx}{x} = S - \tfrac{1}{2}\left(\frac{1}{100} + \frac{1}{200}\right) + \frac{1}{12}\left(\frac{1}{200^2} - \frac{1}{100^2}\right) - \frac{6}{720}\left(\frac{1}{200^4} - \frac{1}{100^4}\right) + \cdots,$$

$$S = \log 2 + 7.5 \times 10^{-3} + 6.25 \times 10^{-6} - 7.8125 \times 10^{-11},$$

$$= 0.69314\ 71806 + 0.00750\ 00000 + 0.00000\ 62500$$

$$- 0.00000\ 00001,$$

$$= 0.70065\ 34305.$$

PROBLEMS.

1. Use the Table of Example 1 to evaluate $\displaystyle\int_0^1 \cos x^2\, dx$. *Ans.* 0.90452.

2. Find the value of the integral $\displaystyle\int_0^{1\cdot2} e^{-x^2}\, dx$, making use of the following:

x	e^{-x^2}	x	e^{-x^2}
0.0	1.0000	0.8	0.5273
0.2	0.9608	1.0	0.3679
0.4	0.8521	1.2	0.2369
0.6	0.6977	Total	4.6427

3. Use the following table to evaluate the integral $\displaystyle\int_0^1 \frac{\sin \pi x}{\pi x}\, dx$:

x	$\dfrac{\sin \pi x}{\pi x}$	x	$\dfrac{\sin \pi x}{\pi x}$
0.0	1.0000	0.6	0.5046
0.2	0.9355	0.8	0.2339
0.4	0.7568	1.0	0.0000
		Total	3.4308

Ans. 0.5895.

4. Use the Euler-Maclaurin formula to prove the following:
$$1^3 + 2^3 + 3^3 + \cdots + n^3 = [\tfrac{1}{2}n\,(n+1)\,]^2.$$

5. Find the value of the following sum:
$$S = \frac{1}{25^2} + \frac{1}{26^2} + \frac{1}{27^2} + \cdots + \frac{1}{50^2}.$$

Ans. 0.021009.

9. Summary and Review. This chapter has been devoted to the development of the elements of the calculus of finite differences. Particular attention has been given to the problem of showing the relationship between the formulas of this calculus and the calculus of infinitesimals.

Following the fundamental definitions and the introduction of factorial symbols, formulas were established for computing the difference $\Delta f(x)$ and a table of such differences for particular functions was found. The theory was extended to differences of higher order and an immediate application was made to the development of the Gregory-Newton interpolation formula, which is the analogue of Taylor's series in the differential calculus. The use of differences of zero in applying the difference calculus to ordinary polynomials was illustrated.

The inverse difference $\Delta^{-1}f(x)$ was identified with the symbol of summation, namely $\sum$, and the elements of the summation calculus were given. A table of summations was developed for special functions and application made to a number of problems.

The relationship between the calculus of finite differences and the infinitesimal calculus was shown by certain operational devices relating D with Δ and $D^{-1} = \int$ with $\Delta^{-1} = \sum$. The chapter concluded with a statement of the Euler-Maclaurin formula for numerical integration and examples were given showing its application both to the evaluation of integrals and the summing of series.

The following problems will serve as a review of the contents of this chapter.

PROBLEMS.

Compute the differences of the following functions:

1. $\dfrac{1}{x(x+1)} - \dfrac{1}{(x+1)(x+2)}$. $Ans. -6(x-1)^{(-4)}$.   2. $\cos^2 x$.

3. $\cos \frac{1}{2}\pi x + \sin \frac{1}{2}\pi x$. $Ans. -2 \sin \frac{1}{2}\pi x$.   4. $\dfrac{x^{(2)} + x}{x^{(2)} - x}$.

5. $x^{(2)} x^{(-2)}$. $Ans. (4x)x^{(-3)}$.   6. $e^x - e^{-x}$.

7. Show that $\Delta x^{-(2)} - \Delta \dfrac{1}{(x+1)^{(2)}} = 6(x-1)^{(-4)}$.

8. If $y = x^{(-n)}$ and $z = (x+1)^{(-n)}$, show that $\Delta(y-z) = -n(n+1)x^{(-n-2)}$.

Verify the following summations:

9. $\sum 5^x (2x+4) = 5^x(\frac{1}{2}x + \frac{3}{8}) + C$.

10. $\sum x\, a^x = \left[\dfrac{x}{(a-1)} - \dfrac{a}{(a-1)^2}\right]a^x + C$.

11. $\sum \cos^2 x = \frac{1}{2}x + \frac{1}{4}\dfrac{\sin(2x-1)}{\sin 1} + C$.

12. $\sum (x^2 - 1) = \frac{1}{6}(2x^3 - 3x^2 - 5x) + C$.

13. $\sum \dfrac{1}{(x+3)(x+4)(x+5)} = -\frac{1}{2}(x+2)^{(-2)} + C$.

14. $\sum (x+3)(x+4) = \frac{1}{3}(x+4)^{(3)} + C$.

15. $\sum a^x \sin \frac{1}{2}\pi x = -\dfrac{a^x}{a^2 + 1}(a \cos \frac{1}{2}\pi x + \sin \frac{1}{2}\pi x) + C$.

16. $\sum \dfrac{x-1}{x(x+1)\,(x+2)} = \dfrac{-2x+1}{2x(x+1)} + C.$

Compute the second differences of the following functions:

17. $\dfrac{1}{(x+3)\,(x+2)\,(x+1)}$.   *Ans.* $12\,x^{(-5)}$.   18. $(x+3)\,(x+2)\,(x+1)$.

19. $\cos^2 x$.   *Ans.* $-2\sin^2 1\cos(2x+2)$.   20. $\dfrac{x+1}{x-1}$.

CHAPTER 2

THE GAMMA AND PSI FUNCTIONS

1. *The Gamma Function.* In order to extend the number of functions that can be summed, we shall now consider the properties and applications of the Gamma function, $\Gamma(x)$, and the closely related Psi function, $\Psi(x)$.

The first of these was originally defined by Leonhard Euler. In modern notation, for values of x such that the real part of the variable is greater than zero, that is, for $R(x) > 0$, the Gamma function is defined by the following integral:

$$\Gamma(x) = \int_0^\infty e^{-t} t^{x-1} \, dt. \tag{1}$$

It is readily shown by the ordinary methods of the calculus, that for integral values of x, that is, when $x = n$ (an integer), we have

$$\Gamma(n) = (n-1)!$$

In other words, the Gamma function is a generalization of the factorial symbol.

Integrating (1) by parts, we can also show that the Gamma function satisfies the following difference equation:

$$\Gamma(x + 1) = x \, \Gamma(x). \tag{2}$$

This equation now provides us with a method for extending the definition of $\Gamma(x)$ to values other than those given by the integral in (1). Thus, as $x \to 0$, we have

$$\lim_{x \to 0} \Gamma(x + 1) = \Gamma(1) = \lim_{x \to 0} x \, \Gamma(x) = 0 \times \Gamma(0);$$

from which we then deduce that, $\Gamma(0) = \infty$.

Moreover, replacing x by $-x$, we have

$$-x \, \Gamma(-x) = \Gamma(1 - x).$$

Thus, when x lies between 0 and 1, the right hand member is defined by (1), and from this we can now define $\Gamma(-x)$ in the interval $-1 < x < 0$. At $x = -1$, $\Gamma(x)$ is now $-\infty$.

This method of extension can be continued into the interval $-2 < x < -1$ from the formula:

$$\Gamma(2 - x) = (1 - x) \, \Gamma(1 - x), \tag{3}$$

and so on throughout the negative x-axis.

THE GAMMA AND PSI FUNCTIONS

The graphical representation of $\Gamma(x)$ along the axis of reals is shown in Figure 1.

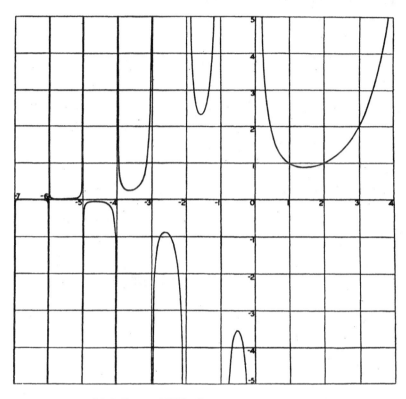

FIGURE 1. THE GAMMA FUNCTION

2. *Properties of the Gamma Function.* The Gamma function has a number of properties, which are proved, some easily and some with difficulty, in advanced treatises. It will be sufficient for our purposes to record a few of these as follows:

(A) $\qquad \Gamma(x)\, \Gamma(1 - x) = \dfrac{\pi}{\sin \pi x}.$ $\qquad\qquad$ (1)

(B) $\qquad \Gamma(x)\, \Gamma(-x) = \dfrac{-\pi}{x \sin \pi x}.$ $\qquad\qquad$ (2)

(C) $\qquad \Gamma(x)\, \Gamma\!\left(x + \dfrac{1}{n}\right) \Gamma\!\left(x + \dfrac{2}{n}\right) \cdots \Gamma\!\left[x + \dfrac{(n-1)}{n}\right]$

$$= (2\pi)^{\frac{1}{2}(n-1)}\, n^{\frac{1}{2}-nx}\, \Gamma(nx), \qquad (3)$$

where n is an integer.

(D) $\qquad 2^{2x}\, \Gamma(x)\, \Gamma\,(x + \tfrac{1}{2}) = 2\sqrt{\pi}\, \Gamma(2x).$ $\qquad\qquad$ (4)

(E) The Gamma function has the following asymptotic expansion:

$\Gamma(x + 1) \sim x^x\, e^{-x}\, \sqrt{2\pi x}\, \times$

$$\left[1 + \frac{1}{12x} + \frac{1}{288x^2} - \frac{139}{51840x^3} - \cdots\right]. \qquad (5)$$

The *Stirling approximation* for $n!$ is a special case of this formula, namely

$$n! \sim n^n\, e^{-n}\, \sqrt{2\pi n}. \qquad (6)$$

(F) $\qquad \dfrac{1}{\Gamma(x)} = x\, e^{\gamma x} \displaystyle\prod_{n=1}^{\infty} \left(1 + \frac{x}{n}\right) e^{-x/n}\,,$ $\qquad\qquad$ (7)

where $\gamma = 0.57721\ 56649\ 01533\ \ldots$ is *Euler's constant.*

(G) $\qquad$ Special values:

$$\Gamma(\tfrac{1}{2}) = \sqrt{\pi},\ \ 1/\Gamma\,(-n) = 0,\, n = 0, 1, 2, \cdots \qquad (8)$$

(H) The *Beta function,* defined by the integral,

$$B(p, q) = \int_0^1 s^{p-1}(1 - s)^{q-1} ds, \qquad (9)$$

has the following representation in terms of the Gamma function:

$$B(p, q) = \frac{\Gamma(p)\, \Gamma(q)}{\Gamma(p + q)}. \qquad (10)$$

3. *Generalization of Factorial x.* By means of the Gamma function we can now make a useful generalization of the factorial symbol $x^{(n)}$. We thus write

$$x^{(n)} = \frac{\Gamma(x + 1)}{\Gamma(x - n + 1)}. \tag{1}$$

This formulation has several advantages. In the first place the original definition for n an integer, namely,

$$x^{(n)} = x \, (x - 1) \, (x - 2) \cdots (x - n + 1), \tag{2}$$

is included as a special case. This is at once evident from the identity

$$\frac{\Gamma(x + 1)}{\Gamma(x - n + 1)}$$

$$= \frac{x \, (x - 1) \, (x - 2) \cdots (x - n + 1) \, \Gamma \, (x - n + 1)}{\Gamma(x - n + 1)}. \tag{3}$$

A second advantage is found in the fact that the formula for $x^{(-n)}$, n an integer, namely,

$$x^{(-n)} = \frac{1}{(n + 1) \, (n + 2) \cdots (x + n)}, \tag{4}$$

is similarly included as a special case. This follows from the following identity:

$$\Gamma(x + n + 1) = (x + n) \, (x + n - 1) \cdots$$
$$(x + 2) \, (x + 1) \, \Gamma \, (x + 1). \tag{5}$$

But the greatest advantage of the generalization comes from the enlarged domain of the values for which the symbol is now defined. Since the Gamma function exists for all values of x in the complex plane, except at the isolated singularities which we have observed in Section 1, we can now extend x and n in the factorial symbol to these same values.

For example, we can now write

$$x^{(0)} = 1, \quad 1^{(n)} = \frac{\Gamma(2)}{\Gamma(2 - n)} = \frac{\sin n\pi}{\pi \, (1 - n)} \, \Gamma(n). \tag{6}$$

Similarly, meaning can be given to the symbol $x^{(\frac{1}{2})}$. Thus, from (1) we have

$$x^{(\frac{1}{2})} = \frac{\Gamma(x + 1)}{\Gamma(x + \frac{1}{2})},$$

and from (4) of Section 2,

$$x^{(\frac{1}{2})} = \frac{\Gamma(x+1) \; 2^{2x} \; \Gamma(x)}{2 \; \sqrt{\pi} \; \Gamma(2x)}.$$

Replacing $\Gamma(x + 1)$ by $x \; \Gamma(x)$ and observing that $B(x, x) = \Gamma^2(x)/\Gamma(2x)$, we finally obtain:

$$x^{(\frac{1}{2})} = \frac{1}{\sqrt{\pi}} \; x \; 2^{2x-1} \; B(x, x). \tag{7}$$

The formulas which were given in Chapter 1 for the difference and the sum of $x^{(n)}$ can now be replaced by the following more general ones:

$$\Delta \; x^{(n)} = n \; \frac{\Gamma(x+1)}{\Gamma(x-n+2)}, \tag{8}$$

$$\sum x^{(n)} = \frac{1}{n+1} \; \frac{\Gamma(x+1)}{\Gamma(x-n)}, \; n \neq -1. \tag{9}$$

These formulas are valid for all values of x and n, and of course include now the case where n is negative.

PROBLEMS

1. Show that $\Gamma(x)$, as defined by the integral (1), Section 1, satisfies the equation:

$$\Gamma(x + 1) = x \; \Gamma(x).$$

Hint: Integrate once by parts.

2. Approximate $\log_{10} 10 ! = 6.559763$ by means of Stirling's formula. Correct this estimate by means of formula (5), Section 1.

3. By means of the transformation $t = s^2$ reduce the integral

$$\int_0^\infty e^{-t} \frac{dt}{(t)^{1/2}}.$$

to the probability integral and from its value show that $\Gamma(\frac{1}{2}) = \sqrt{\pi}$.

4. Noting that $x^{\frac{1}{2}} = \sqrt{x}$, observe that $x^{(\frac{1}{2})}$ tends toward $x^{\frac{1}{2}}$ as x increases by computing the values of each function for $x = 1, 2, 3, 4,$ and 10.

5. Show that

$$B(p, p) = \frac{2 \; \Gamma^2(p+1)}{p \cdot \Gamma(2p+1)}.$$

6. Using the formula of Problem 5, replace $\Gamma(p + 1)$ and $\Gamma(2p + 1)$ by their respective asymptotic values, using (5), Section 2, and thus find that

$$B(p, p) \sim \frac{2 \sqrt{\pi p}}{p \, 2^{2p}} \left(1 + \frac{1}{8p} + \frac{1}{128p^2} + \cdots\right).$$

7. From the results of Problem 6, show that

$$x^{(\frac{1}{2})} \sim \sqrt{x} \left(1 + \frac{1}{8x} + \frac{1}{128x^2} + \cdots\right).$$

Show also that $9^{(\frac{1}{2})} = (65536/12155 \sqrt{\pi}) = 3.04194$ and compare this value with its asymptotic value as given by the formula just derived.

8. Show that

$$x^{(-\frac{1}{2})} = \frac{2}{2x + 1} x^{(\frac{1}{2})}.$$

4. The Psi Function. We must now examine the exceptional case of the summation of $x^{(n)}$ when $n = -1$. That is to say, we must now find a function $u(x)$ such that

$$\sum x^{(-1)} = \sum \frac{1}{x + 1} = u(x).$$

This is equivalent, of course to solving the difference equation

$$\Delta\, u(x) = u(x + 1) - u(x) = \frac{1}{x + 1}. \tag{1}$$

To find a solution of this equation we first differentiate

$$\Gamma(x + 1) = x\, \Gamma(x), \tag{2}$$

from which we get

$$\Gamma'(x + 1) = \Gamma(x) + x\, \Gamma'(x). \tag{3}$$

Dividing the left hand member by $\Gamma(x + 1)$ and the right hand member by $x\, \Gamma(x)$, we have

$$\frac{\Gamma'(x + 1)}{\Gamma(x + 1)} - \frac{\Gamma'(x)}{\Gamma(x)} = \frac{1}{x}. \tag{4}$$

Introducing the Psi function, denoted by $\Psi(x)$ and defined as follows:

$$\Psi(x) = \frac{\Gamma'(x)}{\Gamma(x)}, \tag{5}$$

we now see that the solution of (1) is $\Psi(x + 1)$. We thus obtain the
formula:

$$\sum x^{(-1)} = \sum \frac{1}{x+1} = \Psi(x + 1). \tag{6}$$

An obvious integral representation for $\Psi(x)$ is obtained by taking
the derivative of the integral defining $\Gamma(x)$, [equation (1) of Section 1.]
We thus have

$$\Psi(x) = \frac{1}{\Gamma(x)} \int_0^\infty e^{-t} t^{x-1} \log t \, dt, \quad R(x) > 0, \tag{7}$$

where $R(x)$ means the real part of x.

Other representations are possible such as the following, all of
which are defined for values of x such that the real part of the variable
is positive:

$$\Psi(x) = \int_0^\infty [e^{-t} - (1 + t)^{-x}] \frac{dt}{t}, \tag{8}$$

$$= \log x + \int_0^\infty e^{-tx} \left[\frac{e^t}{1 - e^t} + \frac{1}{t} \right] dt;$$

$$\Psi(x) = -\gamma + \int_0^1 \frac{1 - t^{x-1}}{1 - t} \, dt, \tag{9}$$

$$= -\gamma + x \int_0^1 t^{x-1} \log \left(\frac{t}{1 - t} \right) dt,$$

where γ is Euler's constant.

As in the case of the Gamma function, the difference equation

$$\Psi(x + 1) - \Psi(x) = \frac{1}{x}, \tag{10}$$

provides a method by means of which we can extend the definition
of the Psi function to values other than those given by the integrals.
Thus, it is not difficult to show that $\psi(x)$ is infinite when x is zero or
any negative integer.

The Psi function has one positive zero and an infinite number of
negative zeros. At each of these the Gamma function has either a
maximum or a minimum value. The positive zero, $x_0 = 1.46163$, lies
between 1 and 2. The negative zeros have the following asymptotic
values:

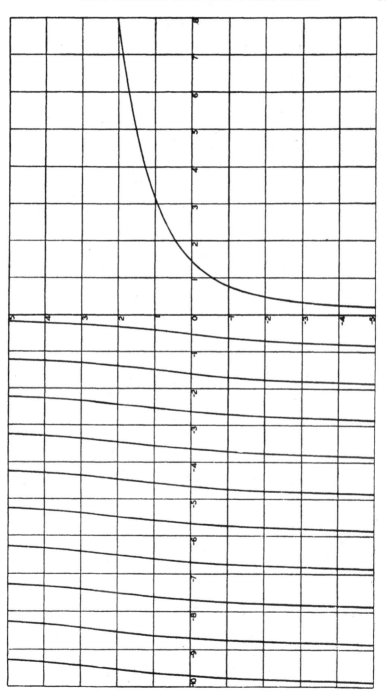

FIGURE 2. THE PSI FUNCTION

$$x_n \sim - n + 1/\log n, \quad n \text{ large.}$$

The graphical representation of $\psi(x)$ over the axis of real values is shown in Figure 2.

5. *Properties of the Psi Function.* A few of the properties of the Psi function and some of its special values will be of interest to us. These are given below without proof, although most of them can be established without too great difficulty from the properties of the Gamma function.

(A) $\Psi(1 - x) = \Psi(x) + \pi \cot \pi x.$ \hfill (1)

(B) $\Psi(-x) = \Psi(x) + \dfrac{1}{x} + \pi \cot \pi x.$ \hfill (2)

(C) $\Psi(nx) = \log n + \dfrac{1}{n} \left\{ \Psi(x) + \Psi\left(x + \dfrac{1}{n}\right) + \Psi\left(x + \dfrac{2}{n}\right) + \right.$

$$\left. \cdots + \Psi\left[x + \dfrac{(n-1)}{n}\right] \right\}, \qquad (3)$$

where n is an integer.

(D) $\Psi(2x) = \log 2 + \tfrac{1}{2} [\Psi(x) + \Psi(x + \tfrac{1}{2})].$ \hfill (4)

(E) The Psi function has the following asymptotic expansion:

$$\Psi(x) \sim \log x - \dfrac{1}{2x} - \sum_{n=1}^{\infty} (-1)^{n-1} \dfrac{B_n}{(2n) \, x^{2n}}. \qquad (5)$$

(F) $\Psi(1 + x) = - \gamma + \displaystyle\sum_{n=1}^{\infty} \dfrac{x}{n \, (n + x)}.$ \hfill (6)

(G) Special values: $\Psi(1) = - \gamma, \; \Psi(\tfrac{1}{2}) = - \gamma - 2 \log 2.$

$$\Psi(\tfrac{1}{3}) = - \gamma - \dfrac{1}{6} \pi \sqrt{3} - \dfrac{3}{2} \log 3;$$

$$\Psi(\tfrac{2}{3}) = - \gamma + \dfrac{1}{6} \pi \sqrt{3} - \dfrac{3}{2} \log 3.$$

$$\Psi(\tfrac{1}{4}) = - \gamma - \tfrac{1}{2}\pi - 3 \log 2;$$

$$\Psi(\tfrac{3}{4}) = - \gamma + \tfrac{1}{2}\pi - 3 \log 2.$$

$$\Psi(\tfrac{1}{5})_| = - \gamma - \tfrac{1}{2}\pi \sqrt{1 + \tfrac{2}{5} \sqrt{5}} - \dfrac{5}{4} \log 5$$

$$- \tfrac{1}{4} \sqrt{5} \log \tfrac{1}{2} (3 + \sqrt{5}), \qquad (7)$$

$$\Psi(\tfrac{2}{5}) = -\gamma - \tfrac{1}{2}\pi \sqrt{1 - \tfrac{2}{5}\sqrt{5}} - \frac{5}{4}\log 5$$
$$+ \tfrac{1}{4}\sqrt{5} \log \tfrac{1}{2}(3 + \sqrt{5}),$$

$$\Psi(\tfrac{3}{5}) = -\gamma + \tfrac{1}{2}\pi \sqrt{1 - \tfrac{2}{5}\sqrt{5}} - \frac{5}{4}\log 5$$
$$+ \tfrac{1}{4}\sqrt{5} \log \tfrac{1}{2}(3 + 2\sqrt{5}),$$

$$\Psi(\tfrac{4}{5}) = -\gamma + \tfrac{1}{2}\pi \sqrt{1 + \tfrac{2}{5}\sqrt{5}} - \frac{5}{4}\log 5$$
$$- \tfrac{1}{4}\sqrt{5} \log \tfrac{1}{2}(3 + 2\sqrt{5}).$$

These values are derived from the following formulas, given originally by K. F. Gauss, in which p and q are integers:

$$\Psi(p/q) = -\gamma - \tfrac{1}{2}\pi \cot (p\pi/q) - \log q + S, \qquad (8)$$
$$\Psi(1 - p/q) = -\gamma + \tfrac{1}{2}\pi \cot (p\pi/q) - \log q + S,$$

where S is defined as follows:

$$S = \sum_{r=1}^{(q-1)/2} \cos (2\pi rp/q) \log [4 \sin^2 (\pi r/q)], \quad (q \text{ odd}),$$

$$S = \sum_{r=1}^{(q-2)/2} \cos (2\pi rp/q) \log [4 \sin^2 (\pi r/q)]$$
$$+ (-1)^p \log 2, \quad (q \text{ even}).$$

$$(9)$$

6. *The Summation of Reciprocal Polynomials.* In Section 6 of Chapter 1 occurs the formula:

$$\sum x^{(-n)} = \frac{x^{(-n+1)}}{-n + 1}. \qquad (1)$$

which is valid except when $n = 1$. The same formula, extended to more general values of n, is given by (9) of Section 3 of this chapter, but with the same exception.

It is now possible, however, to include this case, since we can write

$$\sum x^{(-1)} = \sum \frac{1}{x + 1} = \Psi(x + 1) + C. \qquad (2)$$

Combining this formula with (1), we can now sum the following function:

$$Q(x) = A_1 x^{(-1)} + A_2 x^{(-2)} + \cdots + A_m x^{(-m)}, \qquad (3)$$

where the A_i are arbitrary constants. We thus have

$$\sum Q(x) = A_1 \Psi(x+1) - A_2 x^{(-1)}$$

$$+ \cdots + \frac{1}{-m+1} A_m x^{(-m+1)} + C. \qquad (4)$$

Example 1. Find the sum of the reciprocals of the first hundred integers.

Solution: By means of (2) we have

$$\sum_{x=1}^{100} \frac{1}{x} = \Psi(x) \Big|_1^{101} = \Psi(101) + \gamma.$$

Referring to a table of the Psi function, we thus find

$$\sum_{x=1}^{100} \frac{1}{x} = 4.61016\ 18527 + 0.57721\ 56649 = 5.18737\ 75176.$$

If a table is not available, the value of $\Psi(101)$ can be computed readily by means of formula (5), Section 5. We thus find

$$\Psi(101) = \Psi(100) + \frac{1}{100} = \log 100 + \frac{1}{100} - \frac{1}{200} - \frac{1}{12}\frac{1}{10^4} + \frac{1}{12}\frac{1}{10^9} - \cdots,$$

$$= 4.60517\ 01860 + 01 - .005 - 0.00000\ 83333 = 4.61016\ 18527.$$

More complicated polynomials than those represented by $Q(x)$ can also be summed by the available formulas. We shall consider two cases.

Case I. Let us define the following function:

$$M(x) = \frac{p_n(x)}{(x+1)(x+2)\cdots(x+m)}, \qquad (5)$$

where $p_n(x)$ is an arbitrary polynomial of degree n, which we write as follows:

$$p_n(x) = p_0 + p_1 x + p_2 x^3 + p_3 x^4 + \cdots + p_n x^n. \qquad (6)$$

For convenience n is assumed to be equal to or less than m. This is not a necessary restriction, however, for if n exceeds m then $M(x)$ can be reduced to a function of form (5) plus a polynomial in x. The latter can be summed by the methods previously given.

We now write x^s as the following series:

$$x^s = a_0 + a_1 (x + 1) + a_2 (x + 1) (x + 2) + \cdots$$

$$+ a_s (x + 1) (x + 2) \cdots (x + s), \quad (7)$$

where the coefficients have the following values:

$$a_r = (-1)^{s-r} \frac{\Delta^{r+1} \, 0^{s+1}}{(r + 1) !}, \; r = 0, 1, 2, \cdots, s. \quad (8)$$

For example, we have

$$x^2 = 1 - 3 (x + 1) + (x + 1) (x + 2),$$

$$x^3 = -1 + 7 (x + 1) - 6 (x + 1) (x + 2)$$

$$+ (x + 1) (x + 2) (x + 3),$$

and so on.

Formula (7) is derived from (6) of Section (5), Chapter 1. Thus we have

$$x^{s+1} = \sum_{r=1}^{s+1} x^{(r)} \frac{\Delta^r \, 0^{s+1}}{r !},$$

$$= x + x (x - 1) \frac{\Delta^2 \, 0^{s+1}}{2 !} + x (x - 1) (x - 2) \frac{\Delta^3 \, 0^{s+1}}{3 !}$$

$$+ \cdots + x (x - 1) (x - 2) \cdots (x - s),$$

$$= x \left[1 + \frac{\Delta^2 \, 0^{s+1}}{2 !} (x - 1) + \frac{\Delta^3 \, 0^{s+1}}{3 !} (x - 1) (x - 2) \right.$$

$$\left. + \cdots + (x - 1) (x - 2) \cdots (x - s) \right].$$

Dividing now by x and replacing x by $-x$, we get

$$x^s = (-1)^s \left[1 - \frac{\Delta^2 \, 0^{s+1}}{2 !} (x + 1) + \frac{\Delta^3 \, 0^{s+1}}{3 !} (x + 1) (x + 2) \right.$$

$$\left. + \cdots + (-1)^s (x + 1) (x + 2) \cdots (x + s) \right].$$

When these expansions of x^s have been substituted in (6) and $p_n(x)$ in turn has been substituted in (4), it will be seen that $M(x)$ reduces to a series of the following form:

$$M(x) = A_0 \, x^{(-m)} + A_1 (x + 1)^{(-m+1)}$$

$$+ \cdots + A_n (x + n)^{(-m+n)}. \quad (9)$$

Each term of this series can now be summed and we thus have

$$\sum M(x) = \frac{A_0}{-m+1} x^{(-m+1)} + \frac{A_1}{-m+2} (x+1)^{(-m+2)} + \cdots$$

$$+ \frac{A_n}{-m+n+1} (x+n)^{(-m+n+1)} + C. \quad (10)$$

Example 2. Evaluate the sum

$$S = \sum_{x=0}^{n} \frac{-5x + x^2}{(x+1)(x+2)(x+3)(x+4)}.$$

Solution: Since we can write

$$- 5x + x^2 = - [- 5 + 5(x+1)] + [1 - 3(x+1) + (x+1)(x+2)],$$

we have

$$S = \sum_{x=0}^{n} [6 x^{(-4)} - 8(x+1)^{(-3)} + (x+2)^{(-2)}],$$

$$= \frac{-2}{(x+1)(x+2)(x+3)} + \frac{4}{(x+2)(x+3)} - \frac{1}{x+3} \Big|_0^{n+1},$$

$$= \frac{-n(n+1)}{(n+2)(n+3)(n+4)}.$$

Case II. A more complicated sum of the type under consideration is the following:

$$S = \sum_{x=\alpha}^{\beta} S(x), \quad (11)$$

where we write

$$S(x) = \frac{p_n(x)}{(x+x_1)(x+x_2)(x+x_3) \cdots (x+x_m)}, \quad (12)$$

in which it is assumed (*a*) that $p_n(x)$ is a polynomial of degree n, $n < m$, and (*b*) that the values of x_i differ from one another and are not negative integers.

By means of the theory of partial fractions, we can then write

$$S(x) = \frac{A_1}{x+x_1} + \frac{A_2}{x+x_2} + \frac{A_3}{x+x_3} + \cdots + \frac{A_m}{x+x_m} \quad (13)$$

Making use of the formula

$$\sum \frac{1}{x+r} = \Psi(x+r) + C, \tag{14}$$

we thus obtain

$$S = \sum_{v=1}^{m} A_r \left[\Psi(\beta + 1 + x_r) - \Psi(\alpha + x_r) \right]. \tag{15}$$

Although $S(x)$, as it appears in (13), is the sum of terms of the form $A/(x+a)$, each of which becomes infinite in (11) if $\beta = \infty$, it is nevertheless possible to sum $S(x)$ to infinity provided $n \leqslant m - 2$. In this case we have

$$A_1 + A_2 + \cdots + A_m = 0, \tag{16}$$

and consequently we can write

$$S(x) = - \left[\frac{A_1 x_1}{x(x+x_1)} + \frac{A_2 x_2}{x(x+x_2)} + \cdots + \frac{A_m x_m}{x(x+x_m)} \right]. \tag{17}$$

We now introduce the following expansion of the Psi function:

$$\Psi(1+r) = -\gamma + \sum_{x=1}^{\infty} \frac{r}{x(x+r)}, \tag{18}$$

which has already been recorded as (6) in Section 5.

From this expansion, and observing (16), we thus obtain the following sum:

$$S = \sum_{x=1}^{\infty} S(x) = - \left[A_1 \Psi(1 + x_1) + A_2 \Psi(1 + x_2) \right.$$
$$\left. + \cdots + A_m \Psi(1 + x_m) \right]. \tag{19}$$

Example 3. Evaluate the series:

$$S = \frac{1}{1 \cdot 2 \cdot 3} + \frac{1}{4 \cdot 5 \cdot 6} + \frac{1}{7 \cdot 8 \cdot 9} + \cdots + \frac{1}{3x(3x-1)(3x-2)} + \cdots.$$

Solution: We write

$$S = \frac{1}{27} \sum_{x=1}^{\infty} \frac{1}{x(x-\frac{1}{3})(x-\frac{2}{3})},$$

$$= \frac{1}{6} \sum_{x=1}^{\infty} \left[\frac{1}{x} - \frac{2}{x-\frac{1}{3}} + \frac{1}{x-\frac{2}{3}} \right], \tag{20}$$

$$= -\frac{1}{6} \left[\Psi(1) - 2\Psi(\tfrac{2}{3}) + \Psi(\tfrac{1}{3}) \right].$$

Referring now to the special values of $\Psi(x)$ recorded in (G) of Section 5, we have the following:

$$\Psi(1) = -\gamma, \ \Psi(\tfrac{1}{3}) = -\gamma - \frac{1}{6}\pi\sqrt{3} - \frac{3}{2}\log 3,$$

$$\Psi(\tfrac{2}{3}) = -\gamma + \frac{1}{6}\pi\sqrt{3} - \frac{3}{2}\log 3.$$

When these are substituted in (20), there results

$$S = \frac{\pi}{12}\sqrt{3} - \frac{1}{4}\log 3.$$

PROBLEMS

1. Establish formula (18). First observe that $\Psi(1) = -\gamma$. Then show that $\Psi(1 + r)$ as defined by the formula satisfies the equation

$$\Psi(1 + r) - \Psi(r) = \frac{1}{r}.$$

2. Prove the following:

$$\frac{1}{1 \cdot 1} + \frac{1}{2 \cdot 3} + \frac{1}{3 \cdot 5} + \frac{1}{4 \cdot 7} + \cdots = \log 4.$$

3. Find the value of the following sum:

$$S = \sum_{x=1}^{\infty} \frac{1}{x(3x - 1)}. \quad Ans. \ -\frac{1}{6}\pi\sqrt{3} + \log\sqrt{27}.$$

4. Evaluate:

$$S = \sum_{x=1}^{\infty} \frac{25x + 8}{x(5x + 1)(5x + 4)}.$$

5. Find the value of the sum:

$$S = \sum_{x=1}^{\infty} \frac{2x + 1}{x(3x - 1)(3x - 2)}. \quad Ans. \ \frac{17}{36}\pi\sqrt{3} - \frac{3}{4}\log 3.$$

6. In formula (4) of Section 5, let $x = 1/6$ and hence compute $\Psi(1/6)$. With this value, use formula (1) of Section 5 to find $\Psi(5/6)$. $\quad Ans. \ \Psi(1/6) = -\gamma - \frac{1}{2}\pi\sqrt{3} - \frac{3}{2}\log 3 - 2\log 2.$

7. Making use of the results of Problem 6, show that

$$S = \sum_{x=1}^{\infty} \frac{1}{x(36x^2 - 1)} = -3 + \frac{3}{2}\log 3 + 2\log 2.$$

8. Use formulas (8) and (9) of Section 5 to compute $\Psi(1/8)$ and $\Psi(7/8)$.

9. Prove that

$$\frac{1}{1 \cdot 3} + \frac{1}{3 \cdot 5} + \frac{1}{5 \cdot 7} + \frac{1}{7 \cdot 9} + \cdots = \frac{1}{2}.$$

7. *Polygamma Functions.* In order to extend the theory just given to sums of reciprocal polynomials with repeated factors, it is necessary to introduce some of the properties of polygamma functions.

The term *polygamma function* applies to the Psi function and its derivatives. The *Trigamma* function is $\Psi'(x)$, the *Tetragamma* function is $\Psi''(x)$, the *Pentagamma* function is $\Psi^{(3)}(x)$, the *Hexagamma* function is $\Psi^{(4)}(x)$ and so on.

A few of the properties of these functions, which will be of interest to us are recorded below:

(A) $\quad \Psi^{(n)}(x + 1) = \Psi^{(n)}(x) + (-1)^n \dfrac{n\,!}{x^{n+1}}.$ \hfill (1)

(B) $\quad \Psi^{(n)}(1 - x) + (-1)^{n+1}\, \Psi^{(n)}(x) = (-1)^n\, \pi \dfrac{d^{\,n}}{dx^n} \cot \pi x.$ \hfill (2)

(C) $\quad \Psi^{(n)}(mx) = \dfrac{1}{m^{n+1}} \left\{ \Psi^{(n)}(x) + \Psi^{(n)}\!\left(x + \dfrac{1}{m}\right) + \right.$

$$\Psi^{(n)}\!\left(x + \dfrac{2}{m}\right) + \cdots + \Psi^{(n)}\!\left[x + \dfrac{(m-1)}{m}\right] \Big\}, \ m > 1. \tag{3}$$

(D) $\quad \Psi^{(n)}(x) = (-1)^{n+1}\, n\,! \displaystyle\sum_{r=0}^{\infty} \dfrac{1}{(x + r)^{n+1}}, \ n = 1, 2, \cdots$ \hfill (4)

(E) $\quad \Psi^{(n)}(x) = \displaystyle\int_0^1 \dfrac{t^{x-1} \log^n t}{t - 1}\, dt, \ R(x) > 0.$ \hfill (5)

$$\Psi^{(n)}(x) = (-1)^{n+1} \int_0^{\infty} \log^n (1 + t)\, \dfrac{(1 + t)^{-x}}{t}\, dt,$$

$$R(x) > 0. \tag{6}$$

(F) The polygamma functions, $n > 1$, have the following asymptotic expansion:

$$\Psi^{(n)}(x) \sim (-1)^{n-1} \left[\dfrac{(n - 1)\,!}{x^n} + \dfrac{n\,!}{2x^{n+1}} \right.$$

$$+ \sum_{m=1}^{\infty} (-1)^{m-1} \dfrac{B_m\,(2m + n - 1)\,!}{(2m)\,!\, x^{2m+n}} \Bigg], \tag{7}$$

where the B_m are the Bernoulli numbers.

(G) Special values of the polygamma functions are given in terms of the following series:

$$S_r = 1 + \frac{1}{2^r} + \frac{1}{3^r} + \frac{1}{4^r} + \cdots \; ; \; S_{2n} = 2^{2n-1} \, \pi^{2n} \, B_n/(2n) \, ! \; ;$$

$$T_r = 1 - \frac{1}{3^r} + \frac{1}{5^r} - \frac{1}{7^r} + \cdots \; ; \; T_{2n+1} = E_n \, \pi^{2n+1}/[(2n) \, ! \, 2^{2n+2}]* ;$$

$$\sigma_r = 1 - \frac{1}{2^r} + \frac{1}{4^r} - \frac{1}{5^r} + \frac{1}{7^r} - \frac{1}{8^r} + \frac{1}{10^r} - \frac{1}{11^r} + \cdots .$$

Some of these values are recorded below as follows:

$$\Psi^{(n)}(1) = (-1)^{n+1} \, n \, ! \, S_{n+1}; \; \Psi^{(2n-1)}(1) = (2\pi)^{2n} \, B_n/(4n).$$

$$\Psi^{(n)}(\tfrac{1}{2}) = (-1)^{n+1} \, n \, ! \, (2^{n+1} - 1) \, S_{n+1}, \, n > 0;$$

$$\Psi^{(2n-1)}(\tfrac{1}{2}) = (2\pi)^{2n} \, (2^{2n} - 1) \, B_n/(4n).$$

$$\Psi^{(n)}(\tfrac{1}{4}) = \tfrac{1}{2} \, (-1)^{n+1} \, n \, ! \, 4^{n+1} \left[T_{n+1} + \left(1 - \frac{1}{2^{n+1}} \right) S_{n+1} \right];$$

$$\Psi^{(n)}(\tfrac{3}{4}) = \tfrac{1}{2} \, (-1)^{n+1} \, n \, ! \, 4^{n+1} \left[-T_{n+1} + \left(1 - \frac{1}{2^{n+1}} \right) S_{n+1} \right].$$

$$\Psi^{(n)}(\tfrac{1}{3}) = \tfrac{1}{2} \, (-1)^{n+1} \, n \, ! \, [3^{n+1} \, \sigma_{n+1} + (3^{n+1} - 1) \, S_{n+1}],$$

$$\Psi^{(n)}(\tfrac{2}{3}) = \tfrac{1}{2} \, (-1)^{n+1} \, n \, : \, [-3^{n+1} \, \sigma_{n+1} + (3^{n+1} - 1) \, S_{n+1}].$$

$$\Psi^{(n)}(\infty) = 0, \, n > 0.$$

A few values of S_n and T_n are given in the following table:

n	S_n	n	S_n	n	T_n	n	T_n
1	....	6	1.017343	1	0.785398	6	0.998685
2	1.644934	7	1.008349	2	0.915966	7	0.999555
3	1.202057	8	1.004077	3	0.968946	8	0.999850
4	1.082323	9	1.002008	4	0.988945	9	0.999950
5	1.036928	10	1.000995	5	0.996158	10	0.999983

The graphical representations of $\Psi^{(n)}(x)$, $n = 1, 2, 3, 4$ are shown in Figures 3 and 4.

---

*E_n is the nth *Euler number.* $E_0 = 1$, $E_1 = 1$, $E_2 = 5$, $E_3 = 61$, $E_4 = 1385$.

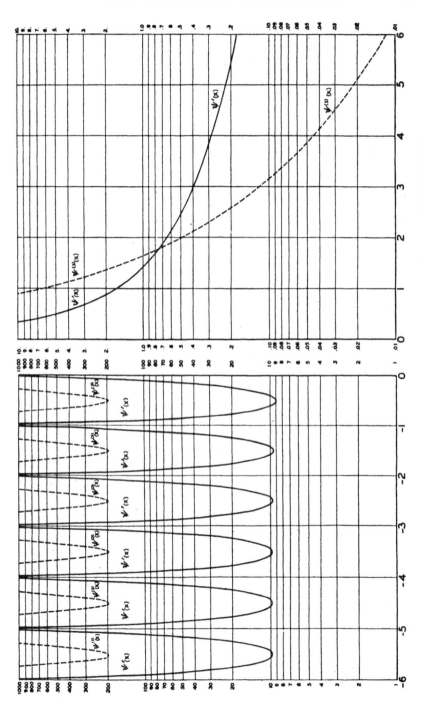

FIGURE 3. THE TRIGAMMA AND PENTAGAMMA FUNCTIONS

8. *The Summation of Reciprocal Polynomials With Repeated Factors.* By means of the polygamma functions it is possible to extend the methods of summation given in Section 6 to the summation of reciprocal polynomials with repeated factors.

We first observe that the difference

$$\Delta\Psi^{(n)}(x) = (-1)^n \frac{n!}{x^{n+1}}, \tag{1}$$

provides us also with the summation formula:

$$\sum \frac{1}{x^n} = \frac{(-1)^{n-1}}{(n-1)!}\Psi^{(n-1)}(x), \quad n = 1, 2, \cdots \tag{2}$$

Example 1. Let us find the value of the sum of the reciprocals of the squares of the first 100 integers.

Solution: Since $n = 2$, we have from (2)

$$\sum_{x=1}^{100}\frac{1}{x^2} = -\Psi'(x)\Big|_{1}^{101} = \Psi'(1) - \Psi'(101).$$

Noting that $\Psi'(1) = S_2 = \pi^2/6 = 1.64493\ 40668$, we need merely to find the value of $\Psi'(101)$ to obtain the desired sum. This can be found either from tables of $\Psi'(x)$ or computed by the asymptotic formula (7) of Section 7. Since $\Psi'(101) = \Psi'(100) - 1/100^2$, we first compute by the asymptotic formula

$$\Psi'(100) \sim \frac{1}{10^2} + \frac{1}{2}\frac{1}{10^4} + \frac{1}{6}\frac{1}{10^6} - \frac{1}{30}\frac{1}{10^{10}} + \cdots$$

$$= 0.01005\ 01666\ 63334.$$

And thus we find

$$\sum_{x=1}^{100}\frac{1}{x^2} = 1.64493\ 40668 - (0.01005\ 01667 - 0.0001),$$

$$= 1.63498\ 39001.$$

Formula (2) can now be extended to include sums of the form

$$S = \sum_{x=a}^{\beta} S(x), \tag{3}$$

where we write

$$S(x) =$$

$$\frac{P(x)}{\prod\limits_{m=1}^{p}(x+x_m)\ \prod\limits_{m=1}^{q}(x+y_m)^2\ \prod\limits_{m=1}^{r}(x+z_m)^3\cdots\prod\limits_{m=1}^{t}(x+u_m)^{n'}}, \tag{4}$$

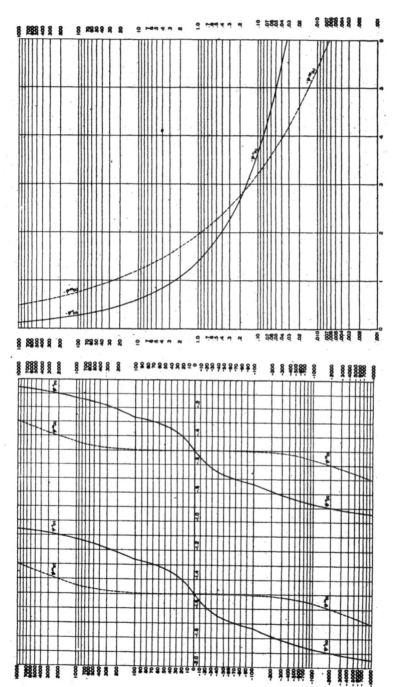

FIGURE 4. THE TETRAGAMMA AND HEXAGAMMA FUNCTIONS

in which $P(x)$ is a polynomial of degree N. It is assumed for convenience that N is less than $p + 2q + 3r + \cdots + nt$. If this is not the case, then $S(x)$ can be written as the sum of a polynomial and a fraction of the desired order.

By the method of partial fractions, $S(x)$ can be expanded as follows:

$$S(x) = \sum_{m=1}^{p} \frac{A_m}{x + x_m} + \sum_{m=1}^{q} \left[\frac{B_{1m}}{x + y_m} + \frac{B_{2m}}{(x + y_m)^2} \right]$$

$$+ \sum_{m=1}^{r} \left[\frac{C_{1m}}{x + z_m} + \frac{C_{2m}}{(x + z_m)^2} + \frac{C_{3m}}{(x + z_m)^3} \right] \qquad (5)$$

$$+ \cdots + \sum_{m=1}^{t} \left[\frac{T_{1m}}{x + u_m} + \frac{T_{2m}}{(x + u_m)^2} + \cdots + \frac{T_{nm}}{(x + u_m)_n} \right],$$

where the numerators of the fractions are constants.

The sum can now be expressed in terms of the coefficients as follows:

$$S = \sum_{m=1}^{p} A_m \left[\Psi(\beta + 1 + x_m) - \Psi(\alpha + x_m) \right]$$

$$+ \sum_{m=1}^{q} B_{1m} \left[\Psi(\beta + 1 + y_m) - \Psi(\alpha + y_m) \right]$$

$$+ \sum_{m=1}^{r} C_{1m} \left[\Psi(\beta + 1 + z_m) - \Psi(\alpha + z_m) \right]$$

$$+ \cdots + \sum_{m=1}^{t} T_{1m} \left[\Psi(\beta + 1 + u_m) - \Psi(\alpha + u_m) \right]$$

$$- \sum_{m=1}^{q} B_{2m} \left[\Psi'(\beta + 1 + y_m) - \Psi'(\alpha + y_m) \right] \qquad (6)$$

$$- \sum_{m=1}^{r} C_{2m} \left[\Psi'(\beta + 1 + z_m) - \Psi'(\alpha + z_m) \right]$$

$$- \cdots - \sum_{m=1}^{t} T_{2m} \left[\Psi'(\beta + 1 + u_m) - \Psi'(\alpha + u_m) \right]$$

$$+ \frac{1}{2!} \sum_{m=1}^{r} C_{3m} \left[\Psi''(\beta + 1 + z_m) - \Psi''(\alpha + z_m) \right] + \cdots$$

$$\pm \frac{1}{(n-1)!} \sum_{m=1}^{t} T_{nm} \left[\Psi^{(n-1)}(\beta + 1 + u_n) - \Psi^{(n-1)}(\alpha + u_n) \right].$$

The case where $\beta = \infty$ is readily included provided we have

$$N \leqslant p + 2q + 3r + \cdots + nt - 2, \tag{7}$$

since series (3) will then be convergent. (See Section 2, Chapter 5).

The sum in this case is obtained by equating to zero all terms in (6) which contain β.

Example 2. Compute the value of the sum

$$S = \sum_{x=1}^{\infty} \frac{5 - 2x}{(x + 3)(x + 1)^3}.$$

Solution: By means of partial fractions, we have

$$\frac{5 - 2x}{(x + 3)(x + 1)^3} = -\frac{11}{8(x+3)} + \frac{11}{8(x+1)} - \frac{11}{4(x+1)^2} + \frac{7}{2(x+1)^3}.$$

Since condition (7) is satisfied, the series converges and we thus obtain from (6) the sum

$$S = \frac{11}{8} \Psi(4) - \frac{11}{8} \Psi(2) - \frac{11}{4} \Psi'(2) - \frac{7}{4} \Psi''(2),$$

$$= \frac{11}{8} \left[1 + \frac{1}{2} + \frac{1}{3} + \Psi(1) \right] - \frac{11}{8} [1 + \Psi(1)]$$

$$- \frac{11}{4} \left[-1 + \Psi'(1) \right] - \frac{7}{4} [2 + \Psi''(1)],$$

$$= \frac{19}{48} - \frac{11}{4} \Psi'(1) - \frac{7}{4} \Psi''(1) = \frac{19}{48} - \frac{11}{4} S_2 + \frac{7}{4} S_3,$$

$$= 0.079464.$$

Example 3. Find the value of the following series:

$$S = \frac{1}{1 \cdot 3^2} + \frac{1}{3 \cdot 10^2} + \frac{1}{5 \cdot 21^2} + \frac{1}{7 \cdot 36^2} + \cdots$$

$$+ \frac{1}{(2x - 1) x^2 (2x + 1)^2} + \cdots$$

Solution: Denoting the general term by $S(x)$, we find the following expansion by the method of partial fractions:

$$S(x) = \frac{2}{x} - \frac{1}{x^2} + \frac{1}{2x - 1} - \frac{5}{2x + 1} - \frac{2}{(2x + 1)^2},$$

$$= \frac{2}{x} - \frac{1}{x^2} + \frac{1}{2}\frac{1}{(x - \frac{1}{2})} - \frac{5}{2}\frac{1}{(x + \frac{1}{2})} - \frac{2}{4}\frac{1}{(x + \frac{1}{2})^2}.$$

From (6) we then obtain the sum

$$S = -2\,\Psi(1) - \frac{1}{2}\Psi(\tfrac{1}{2}) + \frac{5}{2}\Psi\!\left(\frac{3}{2}\right) - \Psi'(1) - \frac{1}{2}\Psi'\!\left(\frac{3}{2}\right).$$

Observing the following values:

$$\Psi(1) = -\gamma,\ \Psi(\tfrac{1}{2}) = -\gamma - 2\log 2,\ \Psi\!\left(\frac{3}{2}\right) = \Psi(\tfrac{1}{2}) + 2,$$

$$\Psi'(1) = S_2 = \pi^2/6,\ \Psi'\!\left(\frac{3}{2}\right) = \Psi'(\tfrac{1}{2}) - 4 = 3\,S_2 - 4 = \tfrac{1}{2}\pi^2 - 4,$$

we at once obtain the desired sum:

$$S = 7 - \frac{5}{12}\pi^2 - \log 16 = 0.115076.$$

PROBLEMS

1. Prove the following:

$$\frac{1}{1^2 \cdot 3^2} + \frac{2}{3^2 \cdot 5^2} + \frac{3}{5^2 \cdot 7^2} + \frac{4}{7^2 \cdot 9^2} + \cdots = \frac{1}{8}.$$

2. Show that

$$\sum_{x=1}^{\infty} \frac{1}{x\,(2x - 1)^2\,(2x + 1)^2} = \frac{3}{2} - 2\log 2.$$

3. Find the value of

$$\sum_{x=1}^{\infty} \frac{1}{x^4}.$$

4. Show that

$$\sum_{x=1}^{\infty} \frac{1}{(2x - 1)^2} = \frac{\pi^2}{8}.$$

CHAPTER 3

OTHER METHODS OF SUMMATION

1. Summation By Differences.—The Gregory Formula. In the applications which we shall make in this chapter, it will be necessary to express the Euler-Maclaurin formula (Section 8, Chapter 1) in terms of differences instead of derivatives.

For this purpose let us now consider the following table of values in which we use the notation:

$$x_n = x + nd, \quad f_n = f(x + nd), \tag{1}$$

where d is the difference interval:

x	$f(x)$	$f(x)$	$\Delta^2 f(x)$	$\Delta^3 f(x)$	$\Delta^4 f(x)$	$\Delta^5 f(x)$	$\Delta^6 f(x)$
x_0	f_0						
x_1	f_1	Δf_0	$\Delta^2 f_0$				
x_2	f_2	Δf_1	$\Delta^2 f_1$	$\Delta^3 f_0$	$\Delta^4 f_0$		
x_3	f_3	Δf_2	$\Delta^2 f_2$	$\Delta^3 f_1$	$\Delta^4 f_1$	$\Delta^5 f_0$	$\Delta^6 f_0$
		Δf_3		$\Delta^3 f_2$		$\Delta^5 f_1$	
....	....	....	....	....	....	....	....
x_{n-3}	f_{n-3}		$\Delta^2 f_{n-4}$		$\Delta^4 f_{n-5}$		$\Delta^6 f_{n-6}$
x_{n-2}	f_{n-2}	Δf_{n-3}	$\Delta^2 f_{n-3}$	$\Delta^3 f_{n-4}$	$\Delta^4 f_{n-4}$	$\Delta^5 f_{n-5}$	
x_{n-1}	f_{n-1}	Δf_{n-2}	$\Delta^2 f_{n-2}$	$\Delta^3 f_{n-3}$			
x_n	f_n	Δf_{n-1}					

The quantities: Δf_0, $\Delta^2 f_0$, and so on, are called the *diagonal*, or *leading*, *differences* associated with the argument x_0. Similarly, the members of the sequence: Δf_{p-1}, $\Delta^2 f_{p-2}$, and so on, are referred to as the *backward differences* belonging to the argument x_p.

Referring now to formula (1), Section 8, Chapter 1, we write the Euler-Maclaurin formula in the following way:

$$\sum_{x=a}^{a+nd} f(x) = \frac{1}{d} \int_a^{a+nd} f(x)\, dx + \tfrac{1}{2}(f_n + f_0) + \frac{1}{12} d(f_n' - f_0') \tag{2}$$

$$- \frac{1}{720} d^3 \left(f_n^{(3)} - f_0^{(3)} \right) + \frac{1}{30240} d^5 \left(f_n^{(5)} - f_0^{(5)} \right) + \cdots,$$

where we use the abbreviations:

$$f_0^{(p)} = f^{(p)}(a), \quad f_n^{(p)} = f^{(p)}(a + nd). \tag{3}$$

The desired formula is obtained by replacing the derivatives of the function by their respective representations in terms of differences. To achieve this transformation we refer to Section 7, Chapter 1 where the derivative symbol D is expressed in the following symbolic formula:

$$D = \frac{1}{d} \log (1 + \Delta) = \frac{1}{d} \left[\Delta - \frac{\Delta^2}{2} + \frac{\Delta^3}{3} - \frac{\Delta^4}{4} + \cdots \right]. \quad (4)$$

Employing the method of symbolic operators, we are able to express similarly the higher derivatives as follows:

$$D^2 = \frac{1}{d^2} \log^2 (1 + \Delta)$$
$$= \frac{1}{d^2} \left[\Delta^2 - \Delta^3 + \frac{11}{12} \Delta^4 - \frac{5}{6} \Delta^5 + \frac{137}{180} \Delta^6 - \cdots \right],$$

$$D^3 = \frac{1}{d^3} \log^3 (1 + \Delta) = \frac{1}{d^3} \left[\Delta^3 - \frac{3}{2} \Delta^4 + \frac{7}{4} \Delta^5 - \frac{15}{8} \Delta^6 + \cdots \right],$$

$$D^4 = \frac{1}{d^4} \log^4 (1 + \Delta) = \frac{1}{d^4} \left[\Delta^4 - 2 \Delta^5 + \frac{17}{6} \Delta^6 - \cdots \right], \quad (5)$$

$$D^5 = \frac{1}{d^5} \log^5 (1 + \Delta) = \frac{1}{d^5} \left[\Delta^5 - \frac{5}{2} \Delta^6 + \cdots \right].$$

The values of f_0', $f_0^{(3)}$, etc. in (2) are then replaced by the following:

$$f_0' = \frac{1}{d} \left[\Delta f_0 - \frac{1}{2} \Delta^2 f_0 + \frac{1}{3} \Delta^3 f_0 - \frac{1}{4} \Delta^4 f_0 \right.$$
$$\left. + \frac{1}{5} \Delta^5 f_0 - \frac{1}{6} \Delta^6 f_0 + \cdots \right], \quad (6)$$

$$f_0^{(3)} = \frac{1}{d^3} \left[\Delta^3 f_0 - \frac{3}{2} \Delta^4 f_0 + \frac{7}{4} \Delta^5 f_0 - \frac{15}{8} \Delta^6 f_0 + \cdots \right].$$

The values of f_n', $f_n^{(3)}$, etc. are similarly expressed in terms of backward differences, but in this case all the signs are positive in the series. We thus write

$$f_n' = \frac{1}{d} \left[\Delta f_{n-1} + \frac{1}{2} \Delta^2 f_{n-2} + \frac{1}{3} \Delta^3 f_{n-3} + \frac{1}{4} \Delta^4 f_{n-4} \right.$$
$$\left. + \frac{1}{5} \Delta^5 f_{n-5} + \frac{1}{6} \Delta^6 f_{n-6} + \cdots \right], \quad (7)$$

$$f_n^{(3)} = \frac{1}{d^3} \left[\Delta^3 f_{n-1} + \frac{3}{2} \Delta^4 f_{n-2} + \frac{7}{4} \Delta^5 f_{n-3} + \frac{15}{8} \Delta^6 f_{n-4} + \cdots \right].$$

When these values are now substituted in (2), we obtain the following formula, called the *Gregory formula*, after James Gregory (1638-1675), who first derived it as a method for integration.

$$\sum_{x=a}^{a+nd} f(x) = \frac{1}{d} \int_a^{a+nd} f(x)\ dx + \frac{1}{2}\ (f_n + f_0) + \frac{1}{12}\ (\Delta f_{n-1} - \Delta f_0)$$

$$+ \frac{1}{24}\ (\Delta f_{n-2}^2 + \Delta^2 f_0) + \frac{19}{720}\ (\Delta^3 f_{n-3} - \Delta f_0^3) \qquad (8)$$

$$+ \frac{3}{160}\ (\Delta^4 f_{n-4} + \Delta^4 f_0) + \frac{863}{60480}\ (\Delta^5 f_{n-5} - \Delta^5 f_0)$$

$$+ \frac{275}{24195}\ (\Delta^6 f_{n-6} + \Delta^6 f_0) + \cdots .$$

The following examples will illustrate the use of this formula in the summation of series:

Example 1. Find the value of the sum

$$S = \sum_{x=1}^{100} x^4,$$

first by the use of formula (2) and then by formula (8).

Solution (a): Setting $f(x) = x^4$, we first compute:

$$f'(x) = 4x^3,\ f''(x) = 12x^2,\ f^{(3)} = 24x,\ f^{(4)} = 24.$$

Since the lower limit of the sum can be 0 as well as 1, we choose the former and thus evaluate the terms in formula (2) at $x = 0$ and $x = 100$. We thus get

$$S = \int_0^{100} x^4\ dx + \frac{1}{2} \times 10^8 + \frac{4}{12} \times 10^6 - \frac{24}{720} \times 10^2,$$

$$= \frac{1}{5} \times 10^{10} + \frac{1}{2} \times 10^8 + \frac{1}{3} \times 10^6 - \frac{10}{3},$$

$$= 2{,}050{,}333{,}330.$$

Solution (b): We first form the following table:

x	x^4	Δ	Δ^2	Δ^3	Δ^4	x	x^4	Δ	Δ^2	Δ^3	Δ^4
0	0					96	849 34656				
1	1	1	14	36		97	885 29281	35 94625	1 12910	2340	
2	16	15	50	60	24	98	922 36816	37 07535	1 15250	2364	24
3	81	65	110			99	960 59601	38 22785	1 17614		
4	256	175				100	1000 00000	39 40399			

Substituting the appropriate values in (8), we have

$$S = \int_0^{100} x^4\ dx + \frac{1}{2} 10^8 + \frac{1}{12}\ (3940399 - 1) + \frac{1}{24}\ (117614 + 14)$$

$$+ \frac{19}{720}\ (2364 - 36) + \frac{3}{160}\ (24 + 24),$$

$$= 20500{,}00000 + 3{,}28366.5 + 4901.16667 + 61.43333 + 0.9$$

$$= 20503{,}33330.$$

Example 2. Given the following table of values of the Trigamma function, $\Psi'(x)$:

x	$\Psi'(x)$	Δ	Δ^2	x	$\Psi'(x)$	Δ	Δ^2
1.00	1.644934			3.96	0.280761		
1.01	1.621214	-23720		3.97	0.286245	-816	
1.02	1.598118	-23096	624	3.98	0.285433	-812	4
1.03	1.575625	-22493	603	3.99	0.284626	-807	5
1.04	1.553712	-21913	580	4.00	0.283823	-803	4

compute the following sum:

$$S = \sum_{x=0}^{300} \Psi'(1 + x/100).$$

Solution: Making appropriate substitution of values in (8), we have

$$S = 100 \int_1^4 \Psi'(x)\,dx + \frac{1}{2}\,(1.644934 + 0.283823)$$

$$+ \frac{1}{12}\,(0.023720 - 0.000803) + \frac{1}{24}\,(0.000624 + 0.000004),$$

Since we have

$$\int_1^4 \Psi'(x)\,dx = \Psi(4) - \Psi(1) = 1 + \frac{1}{2} + \frac{1}{3} = 1.833333 \cdots,$$

the desired sum is at once found to be

$$S = 183.333333 + 0.964379 + 0.001910 + 0.000026,$$

$$= 183.999648.$$

2. *Logarithmic Numbers.** The coefficients in the Gregory formula given in the preceding section are called *logarithmic numbers.* These number, denoted by L_n, are obtained from the following expansion:

$$\frac{x}{\log (1 + x)} = 1 + L_1 x + L_2 x^2 + L_3 x^3 + \cdots + L_n x^n + \cdots, \quad (1)$$

from which we find

$$L_1 = \frac{1}{2},\ L_2 = -\frac{1}{12},\ L_3 = \frac{1}{24},\ L_4 = -\frac{19}{720},$$

$$L_5 = \frac{3}{160},\ L_6 = -\frac{863}{60480} \cdots$$

Their relationship to the Gregory formula is at once evident from the symbolic methods described in Section 7 of Chapter 1. For if we invert the formula

$$dD = \log (1 + \Delta),$$

---

*See H. T. Davis: "The Approximation of Logarithmic Numbers," *American Mathematical Monthly*, Vol. 64, 1957, No. 8, Part II.

we obtain

$$\frac{1}{dD} = \frac{1}{\log{(1 + \Delta)}} = \frac{1}{\Delta} + L_1 + L_2\Delta + L_3\Delta^2 + \cdots \quad (2)$$

Interpreting $1/D$ as an integral and $1/\Delta$ as a sum, and making proper interpretations in their application to $f(x)$, the Gregory formula is obtained.

The logarithmic numbers are also defined by the following integral:

$$L_n = \int_0^1 \frac{s(s-1)\cdots(s-n+1)}{n!}\, ds. \quad (3)$$

This integral can be given a more general form if we write

$$s(s-1)\cdots(s-n+1) = \frac{\Gamma(s+1)}{\Gamma(s-n+1)},$$

$$n! = \Gamma(n+1). \quad (4)$$

When these values are substituted in (3), we then have

$$L_n = \int_0^1 \frac{\Gamma(s+1)\, ds}{\Gamma(n+1)\, \Gamma(s-n+1)}. \quad (5)$$

We next introduce the identity [See (1), Section 2, Chapter 2]

$$\Gamma(x)\, \Gamma(1-x) = \frac{\pi}{\sin{\pi x}}, \quad (6)$$

and use it to write

$$\Gamma(s-n+1) = \Gamma[1-(n-s)] = \frac{\pi}{\sin{\pi(n-s)}\, \Gamma(n-s)},$$

$$= \frac{\pi}{(-1)^{n+1}\sin{\pi s}\, \Gamma(n-s)}. \quad (7)$$

When this value is substituted in (5), we can express L_n in the following useful form:

$$L_n = \frac{(-1)^{n+1}}{\pi} \int_0^1 \frac{\Gamma(s+1)\sin{\pi s}\, \Gamma(n-s)}{\Gamma(n+1)}\, ds. \quad (8)$$

If one observes that the integrand of this integral is the function

$$\sin{\pi s}\, B(s+1, n-s),$$

where $B(p, q)$ is the Beta function defined by (10), Section 2, Chapter 2, then L_n can be expressed as the following double integral:

$$L_n = \frac{(-1)^{n+1}}{\pi} \int_0^1\!\!\int_0^1 \sin{\pi s}\, t^s(1-t)^{n-s-1}\, dt\, ds. \quad (9)$$

The representation of L_n as given by (8) is readily adapted to the evaluation of the constants by means of finite integration. Within the limits of integration $\Gamma(s + 1)$ varies from a minimum value of

$$\gamma_0 = \Gamma(1 + s_0) = 0.88560\ 31943,\ \text{at}\ s_0 = 0.46163\ 21450, \quad (10)$$

to a maximum value of 1 at $s = 0$ and $s = 1$. The ratio

$$\frac{\Gamma(n - s)}{\Gamma(n + 1)}$$

varies from a maximum of $1/n$ at $s = 0$, to a minimum of $1/n\ (n - 1)$ at $s = 1$.

The rapidity of the convergence of the integral is shown from the fact that the values of L_{20} to 9 significant figures and of L_{100} to 7 significant figures were computed by means of formula (8) in which only eleven values of the integrand were used, namely, those over the range $s = 0(.1)1$. These values, to the indicated approximations, are recorded below as follows:

$$L_{20} = -0.00256\ 702255\ \cdots,\ L_{100} = -0.000297\ 4763\ \cdots\ \cdot$$

Equation (8) also provides a means for achieving an interesting asymptotic representation of L_n. To accomplish this we first replace $\Gamma(n - s)$ and $\Gamma(n + 1)$ by their Stirling approximations [See (5), Section 2, Chapter 2,] namely,

$$\Gamma(x) \sim x^{x-1}\ (2\pi x)^{\frac{1}{2}}\ e^{-x}\left[1 + \frac{1}{12x} + \cdots\right]. \quad (11)$$

Making the proper substitutions and simplifying the calculation, we obtain

$$\frac{\Gamma(n - s)}{\Gamma(n + 1)} \sim \frac{e^{-as}}{n}\left(1 - \frac{s}{n}\right)^{-s+\frac{1}{2}} \sim \frac{e^{-as}}{n}, \quad (12)$$

where $a = \log n$.

When this value is substituted in (8), we obtain the following asymptotic expression for L_n:

$$L_n \sim \frac{(-1)^{n+1}}{n\pi}\int_0^1 \Gamma(s + 1)\ \sin\ \pi s\ e^{-as}\ ds. \quad (13)$$

Since the integrand is positive we can now apply to it the law of the mean and thus obtain the following:

$$L_n \sim \frac{(-1)^{n+1}}{n\pi}\ \Gamma(\xi + 1)\int_0^1 \sin\ \pi s\ e^{-as}\ ds,\ 0 \leqslant \xi \leqslant 1, \quad (14)$$

and upon integration,

$$L_n \sim \frac{(-1)^{n+1}\,\Gamma(\xi+1)}{n\,(\log^2 n + \pi^2)} = T_n\,\Gamma(\xi+1), \qquad (15)$$

where we write

$$T_n = \frac{(-1)^{n+1}}{n\,(\log^2 n + \pi^2)}.$$

Using the extreme values of $\Gamma(\xi+1)$, namely 1 and γ_0 given by (10), we obtain the following inequalities for $n = 20$ and $n = 100$:

$$0.002350 < |L_{20}| < 0.002653, \quad 0.0002850 < |L_{100}| < 0.0003218.$$

PROBLEMS

1. Derive the following:

$$\int_0^x x^{(4)}\,dx = \frac{1}{5}\,x^{(5)} + \frac{1}{2}\,x^{(4)} - \frac{1}{3}\,x^{(3)} + \frac{1}{2}\,x^{(2)} - \frac{19}{30}\,x.$$

2. Show that $D^3\,x^{(5)} = 60\,[x^{(2)} - 3x + 7/2]$.

3. Observing that $D \cos x = -\sin x$, use the table of values in Example 1, Section 4 to compute the value of sin 1.

4. Observing that

$$\int_0^x \cos dx = \sin x,$$

use the table of values in Example 1, Section 4 to compute sin 1.

5. To ten decimal places sin 1 = 0.84147 09848. Compare the values found in solving Problems 3 and 4 and explain why one method gave greater accuracy than the other.

6. If L_n is a logarithmic number, prove the following inequality:

$$\frac{n-1}{n+1}\,L_n \leqslant L_{n+1} \leqslant \frac{n}{n+1}\,L_n.$$

3. Lubbock's Summation Formula. If one has computed the sum of values between $x = a$ and $x = b$, but wishes to obtain another sum in which $r - 1$ values are interpolated between each value of the first range, this can be accomplished by means of a formula derived from the Euler-Maclaurin formula [Equation (2), Section 1].

Thus, in this formula, if we subdivide the interval d into r parts, we have the following expansion:

$$\sum_{x=0}^{nr} f(a + xd/r) = \frac{r}{d} \int_a^{a+nd} f(x)\, dx + \frac{1}{2}\, (f_n + f_0)$$

$$+ \frac{d}{12r}\, (f_n' - f_0') - \frac{d^3}{720r^3}\, [f_n^{(3)} - f_0^{(3)}]$$

$$+ \frac{d^5}{30240r^5}\, [f_n^{(5)} - f_0^{(5)}] - \cdots , \qquad (1)$$

where we use the abbreviations:

$$f_0^{(p)} = f^{(p)}(a),\ f_n^{(p)} = f^{(p)}(a + nd).$$

If we now eliminate the integrals between this equation and the Euler-Maclaurin formula [Equation (2), Section 1,] we obtain what is called *Lubbock's summation formula*, published in 1829 by J. W. Lubbock (1803-1865):

$$\sum_{x=0}^{nr} f(a + xd/r) = r \sum_{x=0}^{n} f(a + xd) - \frac{(r - 1)}{2}\, (f_n + f_0)$$

$$- \frac{(r^2 - 1)}{12r}\, d(f_n' - f_0')$$

$$+ \frac{(r^4 - 1)}{720r^3}\, d^3\, [f_n^{(3)} - f_0^{(3)}] \qquad (2)$$

$$- \frac{(r^6 - 1)}{30240r^5}\, d^5\, [f_n^{(5)} - f_0^{(5)}] + \cdots .$$

Example 1. The sum of the fourth powers of the first 10 integers is 25333. Find the sum of the fourth powers of the first 100 numbers.

Solution: In formula (2) we write:

$$a = 0,\ a = 1,\ n = 10,\ f(z) = z^4,\ r = 10.$$

We thus obtain the following equation:

$$\frac{1}{10^4} \sum_{x=0}^{100} x^4 = 10 \sum_{x=0}^{10} x^4 - \frac{9}{2}\, 10^4 - \frac{99}{120} \cdot 4 \cdot 10^3 + \frac{9999}{3000},$$

$$= 253330 - 45000 - 3300 + 3.333,$$

from which we find, $$\sum_{x=0}^{100} x^4 = 2{,}050{,}333{,}330.$$

Example 2. Given the sum

$$S = \sum_{x=0}^{10} \cos \ (x/10) \ = \ 9.1778 \ 47573,$$

find the value of the following:

$$S' = \sum_{x=0}^{50} \cos \ (x/50)$$

Solution: Referring to formula (2) we see that S' is the sum in which four values are interpolated between each of the values in the sum S. Thus we have

$$a = 0, \ d = 1, \ n = 10, \ f(z) = \cos \ (x/10), \ r = 5.$$

Observing that

$$f'(z) = -\frac{1}{10} \sin \ (x/10), \ f^{(3)} = -\frac{1}{10_i^3} \sin \ (x/10), \ \text{etc.},$$

we obtain the following expansion:

$$S' = 5S - 2 \ (\cos 1 + 1) + \left(\frac{24}{12 \cdot 5 \cdot 10} + \frac{624}{720 \cdot 125 \cdot 10^3} \right.$$
$$\left. + \frac{15624}{30240 \cdot 3125 \cdot 10^5} + \cdots \right) \sin 1.$$

Since we find that

$$\sin 1 = 0.84147 \ 09848, \ \cos 1 = 0.54030 \ 23059,$$

we obtain finally

$$S' = 45.8892 \ 37865 - 3.0806 \ 04612 + 0.0336 \ 64675,$$
$$= 42.8422 \ 97928,$$

which has an error of 4 in the last place. This error is not due to the approximation used, but to round-off errors in the multiplications. We shall return again to this problem in the next section.

4. Lubbock's Formula in Terms of Differences. As in the case of the Euler-Maclaurin expansion, the Lubbock summation formula can also be expressed in terms of differences. When the derivatives in (2) of the preceding section are replaced by their equivalent expressions in differences as given in Section 1, the following summation results:

$$\sum_{x=0}^{nr} f(a + xd/r) = r \sum_{x=0}^{n} f(a + xd) - A_1(r) \ (f_n + f_0)$$

$$- A_2(r) \ (\Delta f_{n-1} - \Delta f_0) - A_3(r) \ (\Delta^2 f_{n-2} + \Delta^2 f_0)$$

$$- A_4(r) \ (\Delta^3 f_{n-3} - \Delta^3 f_0) - A_5(r) \ (\Delta^4 f_{n-4} + \Delta^4 f_0) \tag{1}$$

$$- A_6(r) \ (\Delta^5 f_{n-5} - \Delta^5 f_0) - A_7(r) \ (\Delta^6 f_{n-6} + \Delta^6 f_0) - \cdots ,$$

where the coefficients are the following functions:

$$A_1(r) = (r - 1)/2,$$

$$A_2(r) = (r^2 - 1)/(12r),$$

$$A_3(r) = (r^2 - 1)/(24r),$$

$$A_4(r) = (r^2 - 1)(19r^2 - 1)/(720r^3),$$

$$A_5(r) = (r^2 - 1)(9r^2 - 1)/(480r^3), \tag{2}$$

$$A_6(r) = (r^2 - 1)(863r^4 - 145r^2 + 2)/(60480r^5),$$

$$A_7(r) = (r^2 - 1)(275r^4 - 61r^2 + 2)/(24192r^5).$$

The values of these coefficients for a few special values of r are contained in the following table:*

r	$A_2(r)$	$A_3(r)$	$A_4(r)$	r
2	0.125000 000000	0.062500 000000	0.039062 500000	2
4	0.312500 000000	0.156250 000000	0.098632 812500	4
5	0.400000 000000	0.200000 000000	0.126400 000000	5
8	0.656250 000000	0.328125 000000	0.207641 601562	8
10	0.825000 000000	0.412500 000000	0.261112 500000	10
20	1.662500 000000	0.831250 000000	0.526389 062500	20
25	2.080000 000000	1.040000 000000	0.658611 200000	25
30	2.497222 222222	1.248611 111111	0.790740 792182	30
40	3.331250 000000	1.665625 000000	1.054861 132813	40
50	4.165000 000000	2.082500 000000	1.318888 900000	50
60	4.998611 111111	2.499305 555556	1.582870 376800	60
70	5.832142 857143	2.916071 428571	1.846825 400874	70
80	6.665625 000000	3.332812 500000	2.110763 891602	80
90	7.499074 074074	3.749537 037037	2.374691 001905	90
100	8.332500 000000	4.166250 000000	2.638611 112500	100

*A more extensive table for values of r from 2 to 100 by integers will be found in H. T. Davis: *Tables of the Higher Mathematical Functions*, Vol. 2, pp. 251-255.

r	$A_5(r)$	$A_6(r)$	$A_7(r)$	r
2	0.027343 750000	0.020507 812500	0.016113 281250	2
4	0.069824 218750	0.052947 998048	0.042037 963870	4
5	0.089600 000000	0.068032 000000	0.054080 000000	5
8	0.147399 902343	0.112074 851991	0.089208 126073	8
10	0.185418 750000	0.141027 562500	0.112287 656250	10
20	0.373958 593750	0.284550 568360	0.226653 764650	20
25	0.467916 800000	0.356062 986240	0.283629 465600	25
30	0.561805 632718	0.427519 931295	0.340558 958892	30
40	0.749479 199219	0.570350 567078	0.454348 585664	40
50	0.937083 350000	0.713125 680821	0.568091 952053	50
60	1.124652 787423	0.855873 027121	0.681812 181256	60
70	1.312202 387026	0.998604 504447	0.795519 187502	70
80	1.499739 587403	1.141326 062950	0.909217 928370	80
90	1.687268 521377	1.284041 008629	1.022910 263720	90
100	1.874791 668750	1.426751 325184	1.136600 531710	100

Example 1. Given the following table of values of cos x and the sum

$$S = \sum_{x=0}^{10} \cos (x/10) = 9.1778\ 47573,$$

find the value of the following:

$$\sum_{x=0}^{50} \cos (x/50).$$

x	cos x	Δ	Δ^2	Δ^3	Δ^4	Δ^5	Δ^6
0	1.0000 00000						
.1	0.9950 04165	$-$ *49 95835*	$-99\ 41752$				
.2	0.9800 66578	$-149\ 37587$	$-97\ 92502$	*1 49250*	*97846*		
.3	0.9553 36489	$-247\ 30089$	$-95\ 45406$	*2 47096*	*95373*	-2473	
.4	0.9210 60994	$-342\ 75495$	$-92\ 02937$	*3 42469*	*91953*	-3420	-947
.5	0.8775 82562	$-434\ 78432$	$-87\ 68515$	*4 34422*	*87612*	-4341	-921
.6	0.8253 35615	$-522\ 46947$	$-82\ 46481$	*5 22034*	*82397*	-5215	-874
.7	0.7648 42187	$-604\ 93428$	$-76\ 42050$	*6 04431*	*76356*	-6041	-826
.8	0.6967 06709	$-681\ 35478$	$-69\ 61263$	*6 80787*	*69555*	-6801	-760
.9	0.6216 09968	$-750\ 96741$	*$-62\ 10921$*	*7 50342*			
1.0	0.5403 02306	*$-813\ 07662$*					
S = 9.1778 47573							

Solution: Since the problem proposed is that of finding the sum of the values of cos x where four values are interpolated between each of the tabulated values, we make use of expansion (6) in which $r = 5$. Observing that $a = 0$ and $d = 1/10$, we compute as follows:

$$\sum_{x=0}^{50} \cos (x/50) = 5S - 2 \times (1.5403\ 02306) + .4 \times (.0763\ 11827)$$

$$+ .2 \times (.0161\ 52673) - .1264 \times (.0006\ 01092)$$

$$- .0896 \times (.0001\ 67401)$$

$$+ .068032 \times (.0000\ 04328),$$

$$= 42.8422\ 97836.$$

The error in the answer can be determined exactly in this case, since we can write

$$\sum_{x=0}^{50} \cos \frac{x}{50} = \frac{\sin \left(\dfrac{x}{50} - \dfrac{1}{100}\right)}{\sin \left(\dfrac{1}{100}\right)} \Bigg|_{0}^{51} = \tfrac{1}{2} + \frac{\sin\ 1.01}{2 \sin .01},$$

$$= 42.8422\ 97932.$$

The error is thus 96 in the last two places. But this we see is of the same order as the next term in the expansion, namely 92.

Example 2. We shall compute the value of the sum

$$T = \log 100 + \log 101 + \log 102 + \cdots + \log 200,$$

making use of the following table:

x	$\log_{10} x$	Δ	Δ^2	Δ^3	Δ^4	Δ^5	Δ^6
100	2.0000 0000						
		413 9269					
110	2.0413 9269		−36 0413				
		377 8856		5 7767			
120	2.0791 8125		−30 2646		−1 2862		
		347 6210		4 4905		3551	
130	2.1139 4335		−25 7741		−9311		−1137
		321 8469		3 5594		2414	
140	2.1461 2804		−22 2147		−6897		−742
		299 6322		2 8697		1672	
150	2.1760 9126		−19 3450		−5225		−476
		280 2872		2 3472		1196	
160	2.2041 1998		−16 9978		−4029		−325
		263 2894		1 9443		871	
170	2.2304 4892		−15 0535		−3158		−217
		248 2359		1 6285		654	
180	2.2552 7251		−13 4250		−2504		
		234 8109		1 3781			
190	2.2787 5360		−12 0469				
		222 7640					
200	2.3010 3000						
(S)	23.8263 6160						

Solution: We first compute the following values:

$$a_1 = f(200) + f(100) = 4.3010\ 300,$$

$$a_2 = 222\ 7640 - 413\ 9269 = -191\ 1629,$$

$$a_3 = -12\ 0469 - 36\ 0413 = -48\ 0882,$$

$$a_4 = 1\ 3781 - 5\ 7767 = -4\ 3986,$$

$$a_5 = -1\ 2862 - 2504 = -1\ 5366,$$

$$a_6 = 654 - 3551 = -2897,$$

$$a_7 = -217 - 1137 = -1354.$$

We now substitute these values in formula (1) where the coefficients are those corresponding to $r = 10$. We see by inspection that $A_1 = 4.5$ and the other coefficients are obtained from the table. We thus have

$$T = 10S - \sum_{n=1}^{7} A_n(10)\ a_n = 218.9268\ 8453. \tag{3}$$

As in the first example, the error can be exactly determined since

$$\sum \log x = \log \Gamma(x),$$

from which we have

$$T = \log \Gamma(201) - \log \Gamma(100) = 218.9268\ 8499.$$

The error in (3) is thus 46 in the last two places, which is less than one-third of the value of the last term in the approximating series.

This example illustrates one of the difficulties in the use of the Gregory formula for which no easily computed error is available. The magnitude of the error depends, of course, upon the magnitude of the final differences in the table. In the present case the leading differences remain fairly large and tend slowly toward zero.

PROBLEMS

1. Making use of the table in Example 1, Section 1, verify the following sum:

$$S = \sum_{x=1}^{40} x^4 = 217\ 81332.$$

2. Observing that

$$\lim_{r \to \infty} \frac{1}{r} A_n(r) = (-1)^{n-1} L_n,$$

show that the Gregory formula can be derived as the limiting form of the Lubbock formula.

5. *Summation By Means of the Euler-Maclaurin Formula—The Bernoulli Polynomials.* In this and the next section we shall illustrate some of the uses, other than those already exhibited, of the Euler-Maclaurin formula. We shall begin with what we might call the problem of James Bernoulli (1645-1705), who boasted that he was able to compute the sum of the tenth powers of the first thousand integers *intra semi-quadrantem horae* (within $7\frac{1}{2}$ minutes). Whether this celebrated mathematician exaggerated his ability, or whether the hours in his day were longer than in ours, we do not know. But of this we are certain, that the methods as yet available to us in this book are not sufficiently powerful for us to approximate this prodigy of calculation. Thus, if we used the methods explained in Section 6 of Chapter 1, we would first express x^{10} as the sum of factorial terms and then sum each one of these separately. A survey of the values of $\Delta^r 0^{10}/r$! involved in this computation shows that the task would run into hours instead of minutes.

But a much simpler method is available to us if we make use of formula (8), Section 7, Chapter 1, for it we let $f(x) = x^n$, we shall obtain

$$\sum x^n = \int^x x^n \, dx - \tfrac{1}{2}x^n + \frac{B_1}{2!} n \, x^{n-1}$$

$$- \frac{B_2}{4!} n(n-1)(n-2) x^{n-3} + \cdots ,$$

$$= \frac{1}{n+1} \left[x^{n+1} - \tfrac{1}{2}(n+1)x^n + \frac{B_1}{2!}(n+1)n \, x^{n-1} \right. \qquad (1)$$

$$\left. - \frac{B_2}{4!}(n+1)n(n-1)(n-2) x^{n-3} + \cdots \right],$$

We observe that the quantity in the brackets is a polynomial if n is an integer, since the ultimate derivative of x^n is zero. We also observe, that if n is even, the polynomial ends with a term in x, but that when n is odd ($n = 2p + 1$), the last term is a constant and that this constant is $(-1)^p B_{p+1}$.

This polynomial is called a *Bernoulli polynomial* and is denoted by the symbol $B_{n+1}(x)$. The first few of these are recorded below:

$$B_1(x) = x - 1/2,$$

$$B_2(x) = x^2 - x + 1/6,$$

$$B_3(x) = x^3 - 3x^2/2 + x/2,$$

$$B_4(x) = x^4 - 2x^3 + x^2 - 1/30, \tag{2}$$

$$B_5(x) = x^5 - 5x^4/2 + 5x^3/3 - x/6,$$

$$B_6(x) = x^6 - 3x^5 + 5x^4/2 - x^2/2 + 1/42,$$

It is now clear that the summation of x^n between the limits of $x = \alpha$ and $x = \beta$, can be written as follows:

$$S = \sum_{x=\alpha}^{\beta} x^n = \frac{1}{n+1} B_{n+1}(x) \Big|_{\alpha}^{\beta+1}$$

$$= \frac{1}{n+1} [B_{n+1}(\beta+1) - B_{n+1}(\alpha)], \tag{3}$$

and that Bernoulli's problem reduces to the case where $n = 10$, $\alpha = 0, \beta = 1000$.

The sum can be simplified, however, if we write

$$S = 1000^{10} + \sum_{x=0}^{999} x^{10} = 10^{30} + \frac{1}{11} [B_{11}(1000) - B_{11}(0)].$$

This reduces to the comparatively simple sum

$$S = \frac{1}{11} 10^{33} + \tfrac{1}{2} 10^{30} + \frac{5}{6} 10^{27} - 10^{21} + 10^{15} - \tfrac{1}{2} 10^9 + \frac{5}{66} 10^3,$$

$$= 91,409,924,241,424,243,424,241,924,242,500. \tag{4}$$

The Bernoulli polynomials have a number of interesting properties and appear in numerous applications. We shall list some of these properties, most of which are readily proved.

The Bernoulli polynomials can be generated by the following function:

$$\frac{t e^{zt}}{e^t - 1} = \sum_{n=0}^{\infty} \frac{B_n(x) t^n}{n!}, \tag{5}$$

which converges for all values of x and t provided $|t| < 2\pi$. This identity is readily adapted to the proof of some of the following properties:

(A) $\quad B_n(x + 1) - B_n(x) = n x^{n-1}. \tag{6}$

(B) $B_n(1) = B_n(0), n \neq 1, B_1(1) = -B_1(0) = \frac{1}{2}.$ (7)

(C) $B_n(1 - x) = (-1)^n B_n(x).$ (8)

(D) $B_n(kx) = k^{n-1} \sum_{r=0}^{k-1} B_n \left(x + \frac{r}{k} \right).$ (9)

(E) $B_n(x + k) = \sum_{r=0}^{n} {}_nC_r B_{n-r}(x) h^r, {}_nC_r = n!/(n - r)! r!$ (10)

(F) $\sum_{r=0}^{n-1} {}_nC_r B_n(x) = n x^r, n > 0.$ (11)

(G) $\sum_{r=0}^{n-1} {}_nC_r B_r(0) = 0.$ (12)

(H) $B'_n(x) = n B_{n-1}(x).$ (13)

(I) $\int_0^1 B_m(t) B_m(t) dt = (-1)^{m+r} m! n! B_r/(2r)!, m, n > 0,$
$$m + n = 2r,$$

$$= 0, \text{when } m + n \text{ is an odd integer.} \quad (14)$$

(J) $B_n(x) = \sum_{r=0}^{n} {}_nC_r B_{n-r}(0) x^r.$ (15)

It is possible to express the function $x^{(n)}$ in terms of Bernoulli polynomials. To achieve this we first introduce a set of quantities called *differential coefficients of zero*, which are analogous to the differences of zero defined in Section 5 of Chapter 1.

If, in the Maclaurin series,

$$f(p) = f(0) + p Df(0) + \frac{p^2}{2!} D^2 f(0) + \frac{p^3}{3!} D^3 f(0) + \cdots, \quad (16)$$

we let $f(p) = p^{(n)}$, then the coefficient of p^r is the quantity

$$D^r 0^{(n)}/r!, \quad (17)$$

where we define

$$D^r 0^{(n)} = \lim_{x \to 0} D^r x^{(n)}. \tag{18}$$

We shall now call the constants $D^r 0^{(n)}$ the differential coefficients of zero. A few of these quantities, divided by $r\,!$, are given in the following table:

$$\text{Values of } \frac{D^r 0^{(n)}}{r\,!}$$

	$0^{(1)}$	$0^{(2)}$	$0^{(3)}$	$0^{(4)}$	$0^{(5)}$	$0^{(6)}$
$D/1\,!$	1	-1	2	-6	24	-120
$D^2/2\,!$		1	-3	11	-50	274
$D^3/3\,!$			1	-6	35	-225
$D^4/4\,!$				1	-10	85
$D^5/5\,!$					1	-15
$D^6/6\,!$						1
$D^7/7\,!$						
$D^8/8\,!$						
$D^9/9\,!$						
$D^{10}/10\,!$						

	$0^{(7)}$	$0^{(8)}$	$0^{(9)}$	$0^{(10)}$
$D/1\,!$	720	-5040	40302	-362880
$D^2/2\,!$	-1764	13068	-109584	1026576
$D^3/3\,!$	1624	-13132	118124	-1172700
$D^4/4\,!$	-735	6769	-67284	723680
$D^5/5\,!$	175	-1960	22449	-269325
$D^6/6\,!$	-21	322	-4536	63273
$D^7/7\,!$	1	-28	546	-9450
$D^8/8\,!$		1	-36	870
$D^9/9\,!$			1	-45
$D^{10}/10\,!$				1

By means of these values we can now write

$$x^{(n)} = \sum_{r=1}^{n} x^r \frac{D^r 0^{(n)}}{r\,!}, \tag{19}$$

as, for example,

$$x^{(4)} = -6x + 11x^2 - 6x^3 + x^4.$$

Let use now form the sum of (19). We thus have

$$\sum_{x=0}^{x-1} x^{(n)} = \sum_{x=0}^{x-1} \left[x \, D0^{(n)} + x^2 \frac{D^2 0^{(2)}}{2!} + x^3 \frac{D^3 0^{(3)}}{3!} \right.$$

$$\left. + \cdots + x^n \frac{D^n 0^{(n)}}{n!} \right]. \quad (20)$$

Now, observing that the left-hand member is $x^{(n+1)}/(n+1)$, and that

$$\sum_{x=0}^{x-1} x^r = \frac{1}{r+1} [B_{r+1}(x) - B_{r+1}(0)], \quad (21)$$

we obtain the desired expansion:

$$\frac{x^{(n+1)}}{n+1} = \frac{D\,0^{(n)}}{2!} [B_2(x) - B_2(0)] + \frac{D^2 0^{(n)}}{3!} [B_3(x) - B_3(0)]$$

$$+ \frac{D^3 0^{(n)}}{4!} [B_4(x) - B_4(0)] \quad (22)$$

$$+ \cdots + \frac{D^n 0^{(n)}}{(n+1)!} [B_{n+1}(x) - B_{n+1}(0)].$$

Thus, for example, we find that

$$x^{(4)} = 4 \left[(x^2 - x) \doteq (x^3 - \frac{3}{2} x^2 + \frac{1}{2} x) + \tfrac{1}{4}(x^4 - 2x^3 + x) \right],$$

$$= -6x + 11x^2 - 6x^3 + x^4.$$

6. *Euler's Constant and Other Limiting Values.* Euler's constant, denoted by γ, was introduced into a number of formulas in Chapter 2, but its origin and the method of its calculation were not explained. With the Euler-Maclaurin formula available to us, it is now possible to derive this important number.

Euler observed that if $S(n)$ is the sum of the first n terms of the harmonic series, that is,

$$S(n) = 1 + \frac{1}{2} + \frac{1}{3} + \frac{1}{4} + \cdots + \frac{1}{n}, \quad (1)$$

then the difference $S(n) - \log_e n$ approaches a limit as n tends toward infinity. This limit is Euler's number, that is,

$$\gamma = \lim_{n \to \infty} [S(n) - \log_e n] = 0.57721\ 56649\ 01533 \cdots \qquad (2)$$

But the convergence of the quantity within the brackets toward the value of γ is very slow indeed. Thus, if $n = 10{,}000$, we have

$$S(n) = 9.78760\ 60360, \quad \log n = 9.21034\ 03720,$$

whence we have the approximation: $\gamma \sim 0.57726\ 56540$, which is in error in the fifth place.

In order to obtain an adequate formula for computing γ, we make use of the Euler-Maclaurin formulas (2), Section 1, in which we write: $f(x) = 1/x$, $a = 1$, $d = 1$, and replace n by $n - 1$. We thus obtain the following:

$$\sum_{1}^{n} \frac{1}{x} = S(n) = \int_{1}^{n} \frac{1}{x}\, dx + \frac{1}{2n} - \frac{B_1}{2n^2} + \frac{B_2}{4n^4} - \frac{B_3}{6n^6} + \cdots$$

$$+ \left[\tfrac{1}{2} f(1) - \frac{B_1}{2!} f'(1) + \frac{B_2}{4!} f^{(3)}(1) - \frac{B_3}{6!} f^{(5)}(1) + \cdots \right]. \qquad (3)$$

The quantity in brackets is a constant, which we shall call γ. But as we shall soon see, this constant cannot be directly evaluated from the series defining it, since the series is divergent. We shall, however, write (3) as follows:

$$\gamma \sim S(n) - \log n - \frac{1}{2n} + \frac{B_1}{2n^2} - \frac{B_2}{4n^4} + \frac{B_3}{6n^6} - \cdots, \qquad (4)$$

where the symbol "$\sim$" means "asymptotic to" in the sense now to be explained.

Since it can be shown that

$$B_p > \frac{2 \cdot (2p)\,!}{(2\pi)^{2p}}, \qquad (5)$$

the series in (4) is seen to be divergent for all values of n. But the series has the interesting property that, if n is chosen sufficiently large, the sum of a finite number of terms will converge very rapidly to the value of the function which it represents. Such a series is called a *semi-convergent* or *asymptotic* series.

More precisely, we shall say that an asymptotic expansion of a function $f(x)$ is a series of the form,

$$a_0 + a_1/x + a_2/x^2 + \cdots + a_n/x^n + \cdots, \qquad (6)$$

which, although divergent, satisfies the condition that

$$\lim_{|x| \to \infty} x^n \left[f(x) - S_n(x) \right] = 0, \ (n \text{ fixed}) \tag{7}$$

where $S_n(x)$ is the sum $a_0 + a_1/x + a_2/x^2 + \cdots + a_n/x^n$.

In the case of (4), the series diverges at once for $n = 1$, but if $n = 10$, the value of γ is found to ten decimal places by the use of only four terms, and if $n = 10,000$, the same accuracy is obtained with two terms.

A1other problem of similar kind is derived if we let $f(x) = x^{-\alpha}$, $0 < \alpha < 1$. In this case there exists a constant $K(\alpha)$, which is defined by the limit

$$K(\alpha) = \lim_{n \to \infty} \left[\frac{n^{1-\alpha}}{1 - \alpha} - S_\alpha(n) \right], \tag{8}$$

where we write:

$$S_\alpha(n) = 1 + \frac{1}{2^\alpha} + \frac{1}{3^\alpha} + \frac{1}{4^\alpha} + \cdots + \frac{1}{n^\alpha}. \tag{9}$$

The asymptotic representation of the constant is readily shown to be the following:

$$K(\alpha) \sim \frac{n^{1-\alpha}}{1 - \alpha} - S_\alpha(n) + \tfrac{1}{2} n^{-\alpha} - \frac{\alpha}{12} n^{-\alpha-1}$$

$$+ \frac{\alpha(\alpha + 1)(\alpha + 2)}{720} n^{-\alpha-3} \tag{10}$$

$$- \frac{\alpha(\alpha + 1)(\alpha + 2)(\alpha + 3)(\alpha + 4)}{30240} n^{-\alpha-5} + \cdots$$

The case where $\alpha = \tfrac{1}{2}$ is of special interest, since it has been shown by S. Ramanujan that $K(\tfrac{1}{2})$ has the following special form:[*]

$$K(\tfrac{1}{2}) = (1 + (2)^{1/2}) \left[1 - \frac{1}{(2)^{1/2}} + \frac{1}{(3)^{1/2}} - \frac{1}{(4)^{1/2}} + \cdots \right]. \tag{11}$$

In order to evaluate $K(\tfrac{1}{2})$, we substitute appropriate values in (10) and thus obtain the series

$$K(\tfrac{1}{2}) \sim 2(n)^{1/2} - S_{\tfrac{1}{2}}(n) - \frac{1}{2} \frac{1}{(n)^{1/2}} + \frac{1}{24} \frac{1}{n(n)^{1/2}}$$

$$- \frac{1 \cdot 3 \cdot 5}{720 \cdot 8} \frac{1}{n^3(n)^{1/2}} + \frac{1 \cdot 3 \cdot 5 \cdot 7 \cdot 9}{30240 \cdot 32} \frac{1}{n^5(n)^{1/2}} - \cdots \tag{12}$$

---

[*]*Collected Papers*, Cambridge, 1927, p. 48.

If $n = 100$, $S_{\frac{1}{2}}(100) = 18.5896042$ (to 7 decimals), and we approximate

$$K(\tfrac{1}{2}) = 1.4603541 \cdots$$

Since series (12) converges much more slowly than the one used to compute γ, we must start with larger values of n to attain similar accuracy with the same number of terms.

PROBLEMS

1. Show that

$$\Delta x^{(n+1)} = (n + 1)\left[D0^{(n)} x + \frac{D^2 0^{(n)}}{2!} x^2 + \cdots + \frac{D^n 0^{(n)}}{n!} x^n \right].$$

2. Evaluate C_1, C_3, and C_5 from the formula:

$$C_n = 2^{2n+1} \int_0^{\frac{1}{2}} B_n(t)\, dt.$$

Show that the first three terms in the expansion of $\tan x$ are the following:

$$\tan x = - C_1 x + \frac{C_3}{3!} x^3 - \frac{C_5}{5!} x^5 + \cdots$$

3. Compute $R(n)$, where

$$R(n) = S_2(n) - \int_1^n \frac{dx}{x^2},$$

and $S_2(n)$ is defined by (9). If $S_2(10) = 1.54976\,77311$, find $S_2(\infty)$ to ten-place accuracy.

7. *Sums Involving Binomial Coefficients.* Among special sums which are frequently found useful are those which involve binomial coefficients, that is, the function,

$$_nC_r = \frac{n^{(r)}}{r!} = \frac{n!}{(n-r)!\,r!} = \frac{\Gamma(n+1)}{\Gamma(n-r+1)\,\Gamma(r+1)}. \tag{1}$$

Such, for example, is the sum

$$\sum_{x=0}^{n} {}_nC_x^2 = {}_{2n}C_n. \tag{2}$$

We have already encountered sums involving these coefficients in the Gregory-Newton interpolation formula (Section 5, Chapter 1), in series connected with Bernoulli polynomials in Section 5 of this chapter, and elsewhere. They are ubiquitous.

The origin of these quantities is found in the binomial series,

$$(a + b)^n = \sum_{x=0}^{n} a^{n-x} b^x \, {}_nC_x, \tag{3}$$

and many of their properties can be obtained by various manipulations of this sum. Thus, if we write

$$(1 + b)^{2n} = (1 + b)^n (b + 1)^n,$$

$$= \sum_{y=0}^{n} b^n \, {}_nC_y \sum_{z=0}^{n} b^{n-z} \, {}_nC_z,$$

we see that the coefficient of b^n is

$$\sum_{z=0}^{n} {}_nC_z^2;$$

and since the same coefficient in the expansion of $(1 + b)^{2n}$ is ${}_{2n}C_n$, we have established equation (2) above.

An expansion similar to (3), but by no means as well known, is the following:*

$$(a + b)^{(n)} = \sum_{x=0}^{n} a^{(n-x)} b^{(x)} \, {}_nC_x, \tag{4}$$

where the superscripts denote factorials.

The derivation of this formula is very simple, however, for to obtain it we merely refer to the Gregory-Newton series in which we write $f(z) = z^{(n)}$, $x = a$, $p = b$, and $d = 1$.

A transcendental form is given to (4) if we write $a = 1$, $b = z$, and replace the factorial symbols by their equivalent representations in terms of Gamma functions. We thus obtain

$$\frac{\Gamma(2 + z)}{\Gamma(2 + z - n)} \tag{5}$$

$$= \sum_{x=0}^{n} \frac{\Gamma(2)}{\Gamma(2 + x - n)} \frac{\Gamma(z + 1)}{\Gamma(z - x + 1)} \frac{\Gamma(n + 1)}{\Gamma(n - x + 1) \Gamma(x + 1)}.$$

If one assumes that n is not an integer, expansion (5) becomes an infinite series and we thus can write:

---

*This series is given by G. Chrystal: *Algebra*, 1889, Vol. 2, p. 9, where it is attributed to A. T. Vandermonde (1735-1796), although Chrystal says that it was known earlier.

$$\frac{\Gamma(2+z)}{\Gamma(2+z-n)} = \frac{1}{\Gamma(2-n)}\left[1 - \frac{n(n-1)}{(n-1)(n-2)}z\right.$$

$$\left. + \frac{n(n-1)}{(n-2)(n-3)}\frac{z^{(2)}}{2!} - \frac{n(n-1)}{(n-3)(n-4)}\frac{z^{(3)}}{3!}\right. \tag{6}$$

$$\left. + \frac{n(n-1)}{(n-4)(n-5)}\frac{z^{(4)}}{4!} - \cdots\right].$$

As an example of its application, let us write: $z = n = \frac{1}{2}$. We then get the following expansion:

$$\frac{3\pi}{8} = 1 + \frac{1}{1\cdot 3}\frac{1}{2} + \frac{1}{3\cdot 5}\frac{1}{2^2}\frac{1}{2!}$$

$$+ \frac{1\cdot 3}{5\cdot 7}\frac{1}{2^3}\frac{1}{3!} + \frac{1\cdot 3\cdot 5}{7\cdot 9}\frac{1}{2^4}\frac{1}{4!} + \cdots \tag{7}$$

We shall now establish two indentities,* the proofs of which illustrate the methods which are useful in such problems. These identities are the following:

$$\sum_{s=0}^{n-p} {}_{2n+1}C_{2p+2s+1}\;{}_{p+s}C_s = {}_{2n-p}C_p\,2^{2(n-p)}, \tag{8}$$

$$\sum_{s=0}^{n-p} {}_{2n}C_{2p+s}\;{}_{p+s}C_s = \frac{n}{(2n-p)}\,{}_{2n-p}C_p\,2^{2(n-p)}, \tag{9}$$

where n and p are assumed to be integers.

As examples, we shall let $n = 4$, $p = 1$ in (8) and $n = 5$, $p = 2$ in (9). We thus get respectively:

$${}_9C_3\,{}_1C_0 + {}_9C_5\,{}_2C_1 + {}_9C_7\,{}_3C_2 + {}_9C_9\,{}_4C_3$$

$$= 84\cdot 1 + 126\cdot 2 + 36\cdot 3 + 1\cdot 4$$

$$= {}_7C_1\cdot 2^6 = 448.$$

$${}_{10}C_4\,{}_2C_0 + {}_{10}C_6\,{}_3C_1 + {}_{10}C_8\,{}_4C_2 + {}_{10}C_{10}\,{}_5C_3 =$$

$$210\cdot 1 + 210\cdot 3 + 45\cdot 6 + 1\cdot 10 = \frac{5}{8}\,{}_8C_2\cdot 2^6 = 1120.$$

---

*We shall call these the identities of James Moriarty, since we do not know any other source from which such ingenious formulas could have come. See "The Final Problem", *The Memoirs of Sherlock Holmes*.

In order to prove (8) let us first write

$$S = \sum_{s=0}^{n-p} {}_{2n+1}C_{2p+2s+1}\ {}_{p+s}C_s.$$

If we now make the transformation: $s = r - x$, where $r = n - p$, we get

$$S = \sum_{x=0}^{r} {}_{2n+1}C_{2n+1-2x}\ {}_{n-x}C_{r-x},$$

$$= \frac{n!}{r!\,(n-r)!} + \frac{(2n+1)!}{(2n-1)!}\frac{1}{2!}\frac{(n-1)!}{(r-1)!\,(n-r)!}$$

$$+ \frac{(2n+1)!}{(2n-3)!}\frac{1}{4!}\frac{(n-2)!}{(r-2)!\,(n-r)!} +$$

$$\cdots + \frac{(2n+1)!}{(2n+3-2r)!}\frac{1}{(2r-2)!}\frac{(n-r+1)!}{1!\,(n-r)!}$$

$$+ \frac{(2n+1)!}{(2n+1-2r)!}\frac{1}{(2r)!}\frac{(n-r)!}{0!\,(n-r)!},$$

$$= \frac{n!}{(n-r)!\,(2r)!}\left[\frac{(2r)!}{r!} + \frac{2(2n+1)\,(2r)!}{2!\,(r-1)!}\right.$$

$$+ \frac{2^2(2n+1)\,(2n-1)\,(2r)!}{4!\,(r-2)!} +$$

$$\cdots + \frac{2^{r-1}(2n+1)\,(2n-1)\cdots(2n-2r+5)\,(2r)!}{(2r-2)!\,1!}$$

$$\left. + \frac{2^r(2n+1)\,(2n-1)\cdots(2n-2r+3)\,(2r)!}{(2r)!\,0!}\right].$$

Let us now denote the series within the brackets by T and replace $2n + 1$ by $2z$. We then obtain the following:

$$T = \frac{(2r)!}{r!} + \frac{2^2(2r)!}{2!\,(r-1)!}z + \frac{2^4(2r)!}{4!\,(r-2)!}z^{(2)} + \cdots \tag{10}$$

$$+ \frac{2^{2(r-1)}(2r)!}{(2r-2)!\,1!}z^{(r-1)} + \frac{2^{2r}(2r)!}{(2r)!\,0!}z^{(r)}.$$

We shall now show that T is the expansion of the function $2^r F(z)$, where we define:

$$F(z) = (2z + 1)(2z + 3) \cdots (2z + 2r - 1).$$

For this purpose we form the differences of $F(z)$ and thus obtain:

$$\Delta F(z) = (2z + 3)(2z + 5) \cdots (2z + 2r - 1)\, 2r,$$

$$\Delta^2 F(z) = (2z + 5)(2z + 7) \cdots (2z + 2r - 1)\, 2^2 r(r - 1),$$

$$\Delta^s F(z) = (2z + 2s + 1) \cdots (2z + 2r - 1)\, 2^s\, r\,!/(r - s)\,!.$$

Evaluating $\Delta^s F(z)$ at $z = 0$, we obtain

$$\Delta^s F(0) = \frac{(2r)\,!}{(r - s)\,!\, 2^{r-s}\, 1 \cdot 3 \cdot 5 \cdots (2s - 1)}$$

$$= \frac{(2r)\,!\, 2^{2s} s\,!}{(r - s)\,!\, 1 \cdot 2 \cdot 3 \cdots 2s} \frac{1}{2^r}$$

$$= \frac{(2r)\,!\, 2^{2s}\, s\,!}{(r - s)\,!\, (2s)\,!} \frac{1}{2^r}$$

whence we get

$$F(0) = \frac{(2r)\,!}{2^r \cdot r\,!}, \quad \Delta F(0) = \frac{2^2 (2r)\,!}{2^r \cdot 2\,!\, (r - 1)\,!},$$

$$\frac{\Delta^2 F(0)}{2\,!} = \frac{2^4 (2r)\,!}{2^r \cdot 4\,!\, (r - 2)\,!}, \text{ etc.}$$

When these values are substituted in the Gregory-Newton formula (Section 5, Chapter 1), we obtain the expansion $T/2^r$, where T is defined by (10).

The final step in the proof of (8) is merely to observe that the right hand member can be written as follows:

$$_{2n-p}C_p\, 2^{(n-p)} = {}_{n+r}C_{n-r}\, 2^{2r} = \frac{n\,!\, 2^r}{(n - r)\,!\, (2r)\,!}\, F(z), \quad 2z = 2n + 1.$$

The proof of identity (9) follows from a similar argument. The principal difference is the introduction of a variable w instead of z, where $2w = 2n - 1$. The right hand member of (9) then assumes the form

$$\frac{n}{(2n - p)}\, {}_{2n-p}C_p\, 2^{2(n-p)} = \frac{n}{n + r}\, {}_{n+r}C_{n-r}\, 2^{2r}$$

$$= \frac{n\,!\, 2^r}{(n - r)\,!\, (2r)\,!}\, F(w).$$

8. *Moments of the* Bernoulli *Distribution*. In the theory of statistics one encounters what is called the *Bernoulli distribution*, which is defined by the following distribution function:

$$f(x) = {}_nC_x \, q^{n-x} \, p^x, \tag{11}$$

where $p + q = 1$.

It is a matter of some difficulty to compute the moments of this distribution, that is to say, the quantities

$$N_r = \sum_{x=0}^{n} f(x) \, x^r, \; r = 0, 1, 2, \cdots, m,$$

when r is large.

We shall now show that this computation can be reduced to the following sum:

$$N_r = \sum_{s=1}^{r} \frac{\Delta^s 0^r}{s!} \, n^{(s)} \, p^s,$$

where the quantities $\Delta^s 0^r$ are the differences of zero defined in Section 5 of Chapter 1.

We first observe the identity

$$x^{(r)} \, {}_nC_x = n^{(r)} \, {}_{n-r}C_{x-r},$$

from which we then have

$$\sum_{x=0}^{n} x^{(r)} \, {}_nC_x \, q^{n-x} \, p^x = n^{(r)} \, p^r \sum_{x=0}^{n} {}_{n-r}C_{x-r} \, q^{n-x} \, p^{x-r},$$

$$= n^{(r)} \, p^r \sum_{x=r}^{n} {}_{n-r}C_{x-r} \, q^{n-r} \, q^{-x+r} \, p^{x-r},$$

$$= n^{(r)} \, p^r \sum_{s=0}^{n-r} {}_{n-r}C_s \, q^{n-r-s} \, p^s,$$

$$= n^{(r)} \, p^r.$$

Observing the following [identity See (6), Section 5, Chapter 1]:

$$x^r = \sum_{s=1}^{r} \frac{\Delta^s 0^r}{s!} \, x^{(s)},$$

we can now write N_r in the following form:

$$N_r = \sum_{x=0}^{n} {}_nC_x \, q^{n-x} \, p^x \sum_{s=1}^{r} \frac{\Delta^s 0^r}{s!} \, x^{(s)},$$

$$= \sum_{s=1}^{r} \frac{\Delta^s 0^r}{s!} \sum_{x=0}^{n} x^{(s)} \, {}_nC_x \, q^{n-x} \, p^x,$$

$$= \sum_{s=1}^{r} \frac{\Delta^s 0^r}{s!} \, n^{(s)} \, p^s.$$

The first four moments of the Bernoulli distribution are thus readily found to be

$$N_1 = np,$$
$$N_2 = np + n^{(2)}p^2,$$
$$N_3 = np + 3\,n^{(2)}p^2 + n^{(3)}p^3,$$
$$N_4 = np + 7\,n^{(2)}p^2 + 6\,n^{(3)}p^3 + n^{(4)}p^4.$$

PROBLEMS

1. Establish formula (2) by substituting $a = b = n$ in (4).

2. Show that the binomial coefficients satisfy the following equations:

(a) $\displaystyle\sum_{x=0}^{n} {}_nC_x = 2^n.$

(b) $\displaystyle\sum_{x=0}^{n} (-1)^x \, {}_nC_x = 0.$

(c) $\displaystyle\sum_{x=0}^{k} (-1)^x \, {}_nC_x = (-1)^k \, {}_{n-1}C_k.$

3. Assuming the notation

$$S(n) = 1 + \frac{1}{2} + \frac{1}{3} + \frac{1}{4} + \cdots + \frac{1}{n},$$

prove that

$$\sum_{x=1}^{n} (-1)^{x-1} \frac{{}_nC_x}{x} = S(n).$$

4. Establish the following:

$$\sum_{x=0}^{n-2} {}_nC_x \, {}_nC_{x+2} = {}_{2n}C_{n-2}.$$

5. The formula of Leibniz for the nth derivative of the product of two functions u and v is the following:

$$D^n \, (uv) = \sum_{r=0}^{n} D^{n-r}u \, D^r v \, {}_nC_r.$$

Letting $u = x^a$, $v = x^b$, derive formula (2).

6. In the theory of statistics, the following moments are important:

$$M_r = \sum_{x=0}^{n} f(x) \, (x - a)^r,$$

where $f(x)$ is the Bernoulli distribution function defined by (11) and $a = np$.

Verify the following values:

$$M_2 = npq, \; M_3 = npq(q - p), \; M_4 = npq(1 - 6pq + 3npq).$$

CHAPTER 4

SUMMATION BY TABLES

1. *The Tabulation of Sums.* As in the case of integration, the systematic tabulation of finite sums will greatly facilitate the summation of specific series. Such a table is constructed by first differencing a number of functions and then recording the original functions as the sums of the resulting differences. That is to say, one first evaluates $g(x)$ by forming the difference of a given $f(x)$,

$$\Delta f(x) = g(x),$$

from which is then obtained the sum

$$\sum g(x) = f(x).$$

A number of examples of this process have been given in Chapter 1, and the results are recorded in the *Table of Summations* in Section 6 of that chapter. A more extensive list of such sums is given in the *Appendix*. The object of the present chapter is to describe some of the devices used in the construction of the table and to indicate how it can be used.

2. *Summation By Parts.* One of the principal tools used in finding new summation formulas is the formula for summation by parts. This formula, which was introduced in Chapter 1, is the following:

$$\sum u_x \, \Delta v_x = u_x \, v_x - \sum v_{x+1} \, \Delta u_x. \tag{1}$$

We shall now apply it to several examples.

Example 1. We shall evaluate $\sum x \sin (ax + b)$.

Solution: To obtain the sum we write

$$u_x = x, \quad \Delta v_x = \sin (ax + b),$$

from which we have: $\Delta u_x = 1$, $v_x = -\cos (ax + b - \tfrac{1}{2}a)/(2\sin \tfrac{1}{2}a)$.

Substituting these values in (1), we get

$$\sum x \sin (ax + b) = -\frac{x \cos (ax + b - \tfrac{1}{2}a)}{2 \sin \tfrac{1}{2}a} + \sum \frac{\cos (ax + b + \tfrac{1}{2}a)}{2 \sin \tfrac{1}{2}a},$$

$$= -\frac{x \cos (ax + b - \tfrac{1}{2}a)}{2 \sin \tfrac{1}{2}a} + \frac{\sin (ax + b)}{(2 \sin \tfrac{1}{2}a)^2}. \tag{2}$$

The sum which we have just derived can be readily generalized by an iterative formula as follows:

Let us write in (1)

$$u_x = x^{(n)}, \quad \Delta v_x = \sin (ax + b).$$

We then get

$$\sum x^{(n)} \sin (ax + b) = - \frac{x^{(n)} \cos (ax + b - \tfrac{1}{2}a)}{2 \sin \tfrac{1}{2}a}$$
$$+ \frac{n}{2 \sin \tfrac{1}{2}a} \sum x^{(n-1)} \cos (ax + b + \tfrac{1}{2}a). \tag{3}$$

Since the exponent of x has been reduced by 1, it is seen that by successive repetitions of the process the coefficient of the trigonometric function in the last term will ultimately be a constant.

Similarly, one obtains the following complementary formula:

$$\sum x^{(n)} \cos (ax + b) = \frac{x^{(n)} \sin (ax + b - \tfrac{1}{2}a)}{2 \sin \tfrac{1}{2}a}$$
$$- \frac{n}{2 \sin \tfrac{1}{2}a} \sum x^{(n-1)} \sin (ax + b + \tfrac{1}{2}a). \tag{4}$$

Example 2. Establish the following formula:

$$\sum x^{(n)} a^x = \frac{a^x}{a - 1} \left[x^{(n)} - n \frac{a}{a - 1} x^{(n-1)} + n(n - 1) \left(\frac{a}{a - 1} \right)^2 x^{(n-2)} \right.$$
$$\left. - n(n - 1)(n - 2) \left(\frac{a}{a - 1} \right)^3 x^{(n-3)} + \cdots \right]. \tag{5}$$

Solution: To achieve this sum we first use (1), in which we write

$$u_x = x^{(n)}, \quad \Delta v_x = a^x, \quad \text{whence } v_x = a^x/(a - 1).$$

When these values are substituted in (1), we then obtain

$$\sum x^{(n)} a^x = \frac{x^{(n)} a^x}{a - 1} - \frac{na}{a - 1} \sum a^x x^{(n-1)}. \tag{6}$$

Since the second sum has the same form as the first, we can successively repeat the process and formula (5) is the result.

Example 3. Evaluate

$$I = \sum_{x=1}^{n} x^3 \, 2^x.$$

Solution: Since we have

$$x^{(3)} = x + 3 x^{(2)} + x^{(3)},$$

we can now write

$$I = \sum_{x=1}^{n} [x + 3 x^{(2)} + x^{(3)}] \, 2^x. \tag{7}$$

Each term is evaluated by (5) and we thus get

$$\sum x\, 2^x = 2^x(x-2), \quad \sum x^{(2)}\, 2^x = 2^x\,(x^{(2)} - 4x + 8),$$

$$\sum x^{(3)}\, 2^x = 2^x\,(x^{(3)} - 6\, x^{(2)} + 24x - 48).$$

When these sums are substituted in (7), we have

$$I = 2^x\, [x^{(3)} - 3\, x^{(2)} + 13x - 26] \,\Big|_1^{n+1},$$

$$= 2^{n+1}\, (n^3 - 3n^2 + 9n - 13) + 26.$$

Example 4. Find the value of the sum

$$\sum \Psi(x+1).$$

Solution: Letting $u_x = \Psi(x+1)$ and $\Delta v_x = 1$, whence $\Delta u_x = 1/(x+1)$ and $v_x = x$, we get

$$\sum \Psi(x+1) = x\, \Psi(x+1) - x + C. \tag{8}$$

Since $x\, \Psi(x+1) = x\, \Psi(x) + 1$, this formula can be put into the somewhat simpler form

$$\sum \Psi(x+1) = x\, \Psi(x) - x + C'.$$

It will be observed that sum (8) is the analogue of the integral

$$\int \log x\, dx = x \log x - x + C.$$

Example 5. Evaluate

$$\sum x\, \Psi(x+1).$$

Solution: If we let $u_x = \Psi(x+1)$ and $\Delta v_x = x$, we have

$$\sum x\, \Psi(x+1) = \tfrac{1}{2} x^{(2)}\, \Psi(x+1) - \sum \tfrac{1}{2}(x+1)^{(2)}\, \Delta\Psi(x+1),$$

$$= \tfrac{1}{2} x^{(2)}\, \Psi(x+1) - \sum \tfrac{1}{2} x = \tfrac{1}{2} x^{(2)}\, \Psi(x+1) - \tfrac{1}{4} x^{(2)}.$$

This formula is the analogue of the integral

$$\int x \log x\, dx = \tfrac{1}{2} x^2 \log x - \tfrac{1}{4}\, x^2.$$

3. *Formulas Involving Sines and Cosines.* We shall now establish the formulas

$$\sum a^x \cos \theta x = a^x \left\{ \frac{a \cos\,(x-1)\,\theta - \cos \theta x}{a^2 - 2a \cos \theta + 1} \right\}; \tag{1}$$

$$\sum a^x \sin \theta x = a^x \left\{ \frac{a \sin\,(x-1)\,\theta - \sin \theta x}{a^2 - 2a \cos \theta + 1} \right\}. \tag{2}$$

These formulas are the summation equivalents of the following integrals:

$$I = \int e^{ax} \cos \theta x \, dx = e^{ax} \left\{ \frac{a \cos \theta x + \theta \sin \theta x}{a^2 + \theta^2} \right\}; \qquad (3)$$

$$J = \int e^{ax} \sin \theta x \, dx = e^{ax} \left\{ \frac{a \sin \theta x - \theta \cos \theta x}{a^2 + \theta^2} \right\}. \qquad (4)$$

The values of these integrals are customarily obtained by successive integrations by parts. Thus, to obtain (3), we proceed as follows:

$$I = e^{ax} \frac{\sin \theta x}{\theta} - \frac{a}{\theta} \int e^{ax} \sin \theta x \, dx \qquad (5)$$

$$= e^{ax} \frac{\sin \theta x}{\theta} - \frac{a}{\theta} \left(-e^{ax} \frac{\cos \theta x}{\theta} + \frac{a}{\theta} I \right). \qquad (6)$$

We can now solve the second equation for I and thus obtain (3). A similar succession of integrations is used in finding J.

Unfortunately, this simple procedure cannot be used in deriving formulas (1) and (2). But a slight modification of the method leads to the desired result. For it is seen that equation (5) can be written

$$I = e^{ax} \frac{\sin \theta x}{\theta} - \frac{a}{\theta} J,$$

and J can be similarly expresssd as follows:

$$J = -e^{ax} \frac{\cos \theta x}{\theta} + \frac{a}{\theta} I.$$

Although these two equations are equivalent to (6), they may now be regarded as a simultaneous system from which I and J are to be determined. It is this procedure that we now apply to the derivation of (1) and (2).

Thus, denoting by C and S the sums

$$C = \sum a^x \cos \theta x, \quad S = \sum a^x \sin \theta x, \qquad (7)$$

we proceed as follows:

In formula (1) of Section 2, we let $u_x = a^x$ and $\Delta v_x = \cos \theta x$, from which we get

$$u_x = (a - 1) a^x, \quad v_x = \frac{\sin (\theta x - \frac{1}{2}\theta)}{2 \sin \frac{1}{2}\theta}.$$

The formula thus gives us the following:

$$C = a^x \frac{\sin (\theta x - \tfrac{1}{2}\theta)}{2 \sin \tfrac{1}{2}\theta} - \frac{a - 1}{2 \sin \tfrac{1}{2}\theta} \sum a^x \sin (\theta x + \tfrac{1}{2}\theta),$$

$$= a^x \frac{\sin (\theta x - \tfrac{1}{2}\theta)}{2 \sin \tfrac{1}{2}\theta} \tag{8}$$

$$- \frac{a - 1}{2 \sin \tfrac{1}{2}\theta} \sum a^x (\sin \theta x \cos \tfrac{1}{2}\theta + \cos \theta x \sin \tfrac{1}{2}\theta).$$

If we now adopt the following abbreviations:

$$A = a^x \frac{\sin (\theta x - \tfrac{1}{2}\theta)}{2 \sin \tfrac{1}{2}\theta}, \quad \lambda = \frac{a - 1}{2 \sin \tfrac{1}{2}\theta},$$

then (8) can be written

$$C = A - (\lambda \cos \tfrac{1}{2}\theta) S - (\lambda \sin \tfrac{1}{2}\theta) C. \tag{9}$$

Applying a similar technique to S and using the abbreviation,

$$B = a^x \frac{\cos (\theta x - \tfrac{1}{2}\theta)}{2 \sin \tfrac{1}{2}\theta},$$

we obtain the following equation:

$$S = -B + (\lambda \cos \tfrac{1}{2}\theta) C - (\lambda \sin \tfrac{1}{2}\theta) S. \tag{10}$$

Equation (9) and (10) now provide a linear system from which both C and S can be determined in terms of A, B, and λ. After some simplification formulas (1) and (2) are obtained.

4. *Sums of Powers of Sines and Cosines.* The sums of the powers of sines and cosines are readily obtained from the sums of the functions themselves, since the powers can be expressed linearly in terms of the functions with multiple angles. Thus we have

$$\sum \sin^2 \theta x = \sum \tfrac{1}{2}(1 - \cos 2\theta x) = \tfrac{1}{2}x - \tfrac{1}{4} \frac{\sin (2\theta x - \theta)}{\sin \theta}, \tag{1}$$

$$\sum \cos^2 \theta x = \sum \tfrac{1}{2}(1 + \cos 2\theta x) = \tfrac{1}{2}x + \tfrac{1}{4} \frac{\sin (2\theta x - \theta)}{\sin \theta}. \tag{2}$$

Similarly, since we have

$$\sin^3 \theta x = \tfrac{1}{4}(3 \sin \theta x - \sin 3\theta x), \quad \cos^3 \theta x = \tfrac{1}{4}(3 \cos \theta x + \cos 3\theta x),$$

we compute the following sums:

$$\sum \sin^3 \theta x = -\frac{3}{8} \frac{\cos (\theta x - \tfrac{1}{2}\theta)}{\sin \tfrac{1}{2}\theta} + \frac{1}{8} \frac{\cos (3\theta x - \tfrac{3}{2}\theta)}{\sin \tfrac{3}{2}\theta}, \tag{3}$$

$$\sum \cos^3 \theta x = \frac{3}{8} \frac{\sin (\theta x - \frac{1}{2}\theta)}{\sin \frac{1}{2}\theta} + \frac{1}{8} \frac{\sin (3\theta x - \frac{3}{2}\theta)}{\sin \frac{3}{2}\theta}. \tag{4}$$

When the above sums are taken between the limits $x = \theta$ and $x = n - 1$, the following formulas are obtained:

$$\sum_{x=0}^{n-1} \sin^2 \theta x = \tfrac{1}{2}n - \frac{\cos (n - 1) \theta \sin n\theta}{2 \sin \theta}, \tag{5}$$

$$\sum_{x=0}^{n-1} \cos^2 \theta x = \tfrac{1}{2}n + \frac{\cos (n - 1) \theta \sin n\theta}{2 \sin \theta}; $$

$$\sum_{x=0}^{n-1} \sin^3 \theta x = \frac{3}{4} \frac{\sin \frac{1}{2}n\theta \sin \frac{1}{2}(n - 1) \theta}{\sin \frac{1}{2}\theta} \tag{6}$$
$$- \frac{1}{4} \frac{\sin \frac{3}{2}n\theta \sin \frac{3}{2}(n - 1) \theta}{\sin \frac{3}{2}\theta},$$

$$\sum_{x=0}^{n-1} \cos^3 \theta x = \frac{3}{4} \frac{\sin \frac{1}{2}n\theta \cos \frac{1}{2}(n - 1) \theta}{\sin \frac{1}{2}\theta} \tag{7}$$
$$+ \frac{1}{4} \frac{\sin \frac{3}{2}n\theta \cos \frac{3}{2}(n - 1) \theta}{\sin \frac{3}{2}\theta}.$$

If θ is replaced by $2k\pi/n$, where $2k$ is an integer not divisible by n, then we obtain from formulas (5), (6), and (7) the following sums:

$$\sum_{x=0}^{n-1} \sin^2 \frac{2k\pi}{n} x = \sum_{x=0}^{n-1} \cos^2 \frac{2k\pi}{n} x = \tfrac{1}{2}n, \tag{8}$$

$$\sum_{x=0}^{n-1} \sin^3 \frac{2k\pi}{n} x = \sum_{x=0}^{n-1} \cos^3 \frac{2k\pi}{n} x = 0.$$

These formulas are readily extended to higher powers if we observe that $\sin^k z$ and $\cos^k z$, $z = \theta x$, can be written as linear sums of $\sin mz$ and $\cos mz$. Thus we have

$$\sin^{2p} z = \frac{1}{2^{2p-1}} [a_0 - a_1 \cos 2z + a_2 \cos 4z \tag{9}$$
$$- \cdots + (-1)^p a_p \cos 2pz],$$

$$\cos^{2p} z = \frac{1}{2^{2p-1}} [a_0 + a_1 \cos 2z + a_2 \cos 4z + \cdots + a_p \cos 2pz],$$

where $a_0 = \tfrac{1}{2} \, {}_{2p}C_p, \ a_r = {}_{2p}C_{p+r}, \ r \neq 0$.

Similarly, for odd powers of the functions, we have

$$\sin^{2p+1} z = \frac{1}{2^{2p}} [b_0 \sin z - b_1 \sin 3z + \cdots \tag{10}$$
$$+ (-1)^p b_p \sin (2p + 1) z],$$
$$\cos^{2p+1} z = \frac{1}{2^{2p}} [b_0 \cos z + b_1 \cos 3z + \cdots + b_n \cos (2p + 1) z],$$

where $b_r = {}_{2p+1}C_{n+r+1}$.

The problem of evaluating the sums of $\sin^k \theta x$ and $\cos^k \theta x$ is thus reduced to the evaluating the sums of the first powers of the functions.
If we observe the following values:

$$\sum_{x=0}^{n-1} \cos \frac{2k\pi}{n} x = \sum_{x=0}^{n-1} \sin \frac{2k\pi}{n} x = 0,$$

when $2k$ is an integer not divisible by n, then we obtain from (9) and (10) the following sums:

$$\sum_{x=0}^{n-1} \sin^{2p} \frac{2k\pi}{n} x = \sum_{x=0}^{n-1} \cos^{2p} \frac{2k\pi}{n} x = \frac{n}{2^{2p}} {}_{2p}C_p,$$

$$\tag{11}$$

$$\sum_{x=0}^{n-1} \sin^{2p+1} \frac{2k\pi}{n} x = \sum_{x=0}^{n-1} \cos^{2p+1} \frac{2k\pi}{n} x = 0.$$

PROBLEMS

1. Prove that

$$\sum_{x=0}^{n-1} \sin^4 \theta x = \frac{1}{8} \left[3n - 4 \frac{\sin n\theta \cos (n-1) \theta}{\sin \theta} + \frac{\sin 2n\theta \cos 2(n-1) \theta}{\sin 2\theta} \right].$$

2. Prove that

$$\sum_{x=0}^{n-1} \cos^4 \theta x = \sum_{x=0}^{n-1} \sin^4 \theta x + \frac{\sin n\theta \cos (n-1) \theta}{\sin \theta}.$$

3. Given the following sum:

$$S = \sum_{x=0}^{n-1} \sin \frac{2k\pi}{n} x \sin \frac{2m\pi}{n} x,$$

show (a) that $S = 0$, if neither $k - m$ nor $k + m$ is divisible by n, or if both are divisible by n; (b) that $S = \frac{1}{2}n$, if $k - m$ is divisible by n and $k + m$ is not divisible by n; (c) that $S = -\frac{1}{2}n$, if $k - m$ is not divisible by n and $k + m$ is divisible by n.

4. Given the following sum:

$$C = \sum_{x=0}^{n-1} \cos \frac{2k\pi}{n} x \cos \frac{2m\pi}{n} x,$$

show (a) that $C = 0$, if neither $k - m$ nor $k + m$ is divisible by n; (b) that $C = \frac{1}{2}n$, if either $k - m$ or $k + m$ is divisible by n, but not both; (c) that $C = n$, if both $k - m$ and $k + m$ are disivisble by n.

5. *Some Miscellaneous Examples.* By means of the methods previously described various summation formulas have been derived and recorded in the *Appendix*. We shall now illustrate the application of these formulas to a few problems.

Example 1. Find the value of the following series:

$$S = \tan^{-1} \frac{2}{1} + \tan^{-1} \frac{2}{2^2} + \tan^{-1} \frac{2}{3^2} + \tan^{-1} \frac{2}{4^2} + \tan^{-1} \frac{2}{5^2} + \cdots \quad (1)$$

Solution: That the series converges is readily proved by means of the integral test, [See (7), Section 2, Chapter 5] since we have

$$S - \tan^{-1} 2 < \int_1^\infty \tan^{-1} \frac{2}{x^3} \, dx = \frac{1}{2} \int_1^\infty \frac{\tan^{-1} 2u}{u(u)^{1/2}} \, du < \frac{1}{2}\pi.$$

Let us now write S as the following sum:

$$S = \sum_{x=0}^\infty \tan^{-1} \left[\frac{2}{(2x+1)^2} \right] + \sum_{x=1}^\infty \tan^{-1} \left[\frac{2}{(2x)^2} \right]. \quad (2)$$

Consulting the *Table of Finite Sums* in the *Appendix*, we find in $(G, 4)$ the formula:

$$\sum \tan^{-1} \left\{ \frac{\theta}{1 + \theta(h + x\theta) + (h + x\theta)^2} \right\} = \tan^{-1} (h + x\theta).$$

For the special values $\theta = 2$, $h = 0$, and $\theta = 2$, $h = 1$, this formula reduces respectively to the two sums in (2).

We thus obtain

$$S = \tan^{-1} 2x \bigg|_0^\infty + \tan^{-1} (2x - 1) \bigg|_1^\infty = \frac{3\pi}{4}.$$

Example 2. Evaluate the series:

$$S = 1 + \frac{1 \cdot 2}{2 \cdot 3} + \frac{2 \cdot 4}{3 \cdot 4} + \frac{3 \cdot 8}{4 \cdot 5} + \frac{4 \cdot 16}{5 \cdot 6} + \cdots + \frac{n \cdot 2^n}{(n+1)(n+2)}.$$

Solution: Since the series can be written as the following sum:

$$S = 1 + \sum_{x=1}^n \frac{x \cdot 2^x}{(x+1)(x+2)}.$$

the desired value is obtained by the use of formula (C, 27) in the *Table of Finite Sums*. We thus get

$$S = 1 + \frac{2^x}{x+1}\Big|_1^{n+1} = \frac{2^{n+1}}{n+2}.$$

Example 3. Establish the following:

$$1 - \frac{m}{n+1} + \frac{m(m-1)}{(n+1)(n+2)} - \frac{m(m-1)(m-2)}{(n+1)(n+2)(n+3)} + \cdots = \frac{n}{m+n}$$

where n is assumed to be a positive number and exceeds in value $|m|$.

Solution: In order to establish the convergence of the series, we first observe that, if m is an integer, the series terminates, and for any other value of m there exists a term beyond which all the remaining terms are of one sign. Thus we can use the Raabe test in examining the problem of convergence.

If, for simplicity, we now replace m by $-\mu$, then the general term of the series can be written:

$$u(x) = \frac{(\mu+1)(\mu+2)\cdots(\mu+x+1)}{(n+1)(n+2)(n+3)\cdots(n+x)} = \frac{\Gamma(\mu+x)}{\Gamma(\mu)}\frac{\Gamma(n+1)}{\Gamma(n+x+1)}.$$

The Raabe test [See (5), Section 2 of next Chapter] for the convergence of a series, the terms of which are of one sign, makes use of the quantity

$$a(x) = \frac{u(x)}{u(x+1)} - 1,$$

where $u(x)$ is the general term of the series. It asserts that if $\lim\limits_{x\to\infty} a(x) = 0$, but if $\lim\limits_{x\to\infty} x\, a(x) > 1$, the series converges; if $\lim\limits_{x\to\infty} x\, a(x) < 1$, the series diverges; if $\lim\limits_{x\to\infty} x\, a(x) = 1$, the test fails.

In the present case

$$u(x+1) = K\frac{\Gamma(\mu+x+1)}{\Gamma(n+x+2)} = K\frac{(\mu+x)\,\Gamma(\mu+x)}{(n+x+1)\,\Gamma(n+x+1)},$$

where $K = \Gamma(n+1)/\Gamma(\mu)$. Hence we have

$$\lim_{x\to\infty} x\, a(x) = n - \mu + 1,$$

whence series (3) converges since $n > |m| = |\mu|$.

In order to achieve the desired sum, we now refer to formula $(H, 12)$ in the *Table of Finite Sums* from which we get

$$u(x) = \frac{\Gamma(n+1)}{\Gamma(\mu)} \sum_{x=0}^{\infty} \frac{\Gamma(\mu+x)}{\Gamma(n+x+1)} = \frac{\Gamma(n+1)}{\Gamma(\mu)}\frac{1}{(\mu-n)}\frac{\Gamma(\mu+x)}{\Gamma(n+x)}\Big|_0^{\infty},$$

$$= \frac{n}{n-\mu} = \frac{n}{n+m}.$$

6. *The Summation of $\mathrm{Log}_e\ \Gamma(x)$.* A problem which presents certain points of interest is found in the evaluation of the sum of $\log_e \Gamma(x)$, which, by formula $(F, 2)$ in the *Table of Finite Sums*, can be written as follows:

$$\sum_{x=1}^{n} \log \Gamma(x) = n \log \Gamma(n+1) - \sum_{x=1}^{n} x \log x. \qquad (1)$$

The problem is thus reduced to the evaluation of the series

$$L(n) = \sum_{x=1}^{n} x \log x. \qquad (2)$$

For this purpose we make use of the Euler-Maclaurin formula (1), Section 8, Chapter 1, in which we substitute $f(x) = x \log x$ and thus obtain:

$$\sum_{x=1}^{n} x \log x = \int_{1}^{n} x \log x \, dx + \tfrac{1}{2} n \log n + S(n) - S(1), \qquad (3)$$

$$= \left(\tfrac{1}{2} n^2 + \tfrac{1}{2} n + \frac{1}{12} \right) \log n - \tfrac{1}{4} n^2 + \tfrac{1}{4} + S(n) - S(1),$$

where we abbreviate:

$$S(n) = \frac{B_2}{4!} \frac{1}{n^2} - \frac{3! \cdot B_2}{6!} \frac{1}{n^4} + \frac{5! \cdot B_4}{8!} \frac{1}{n^6} - \cdots. \qquad (4)$$

Although series (4) is divergent for all values of n, it is the asymptotic expansion of the integral

$$S(n) = \int_{0}^{\infty} e^{-t} \left[\frac{B_2}{4!} \frac{t}{n^2} - \frac{B_3}{6!} \frac{t^3}{n^4} + \frac{B_4}{8!} \frac{t^5}{n^6} - \cdots \right] dt, \qquad (5)$$

and thus can be used to evaluate $S(n)$ for sufficiently large values of n.

Observing the expansion

$$\frac{t}{e^t - 1} = 1 - \frac{1}{2} t + B_1 \frac{t^2}{2!} - B_2 \frac{t^4}{4!} + B_3 \frac{t^6}{6!} - \cdots, \qquad (6)$$

and making in (5) the transformation: $t = ns$, we can write the integral as follows:

$$S(n) = \int_{0}^{\infty} e^{-ns} \left[1 - \tfrac{1}{2} s + \frac{B_1}{2!} s^2 - \frac{s}{e^s - 1} \right] \frac{ds}{s^3}. \qquad (7)$$

The constant $S(1)$ which occurs in (3) cannot be evaluated from (4), but can be computed with some effort from the integral just given. However, a much easier method is available to us. Since $S(n)$ approaches zero rapidly for large values of n, we see from (3) that

$$\lim_{n \to \infty} [L(n) - P(n)] = K, \tag{8}$$

where $K = \frac{1}{4} - S(1)$ is a constant and where we use the convenient abbreviation:

$$P(n) = \left(\tfrac{1}{2}n^2 + \tfrac{1}{2}n + \frac{1}{12}\right) \log n - \tfrac{1}{4}n^2.$$

But since K is actually equal to $L(n) - P(n) - S(n)$ for all values of n, and since $S(n)$ can be computed with small error from (4) for relatively small values of n, we can attain at least ten-place accuracy in the evaluation of K when $n = 10$, that is,

$$K = L(10) - P(10) - S(10). \tag{9}$$

We thus find by direct calculation:

$$L(10) = 102.08283\ 05519,$$

$$P(10) = 101.83406\ 22058,$$

$$S(10) = \quad 0.00001\ 38691,$$

and hence we compute:

$$K = \quad 0.24875\ 44770.$$

This constant, it will be observed, resembles Euler's constant, which we have previously discussed. It is actually the natural logarithm of what is called *Glaisher's constant*, $A = 1.28242\ 71291$ . . . , which occurs in the theory or products of the form: $1^a\ 2^b\ 3^c$ . . . x^z, where $a = 1^r, b = 2^r, c = 3^r$, etc.

We now see that we can write (1) in the following form:

$$\sum_{x=1}^{n} \log \Gamma(x) = n \log \Gamma(n + 1) - P(n) - S(n) - K. \tag{10}$$

As an example we have

$$\sum_{x=1}^{100} \log \Gamma(x) = 100 \log \Gamma(101) - P(100) - S(100) - K, \tag{11}$$

where, from tables, we find: $100 \log \Gamma(101) = 36373.93756$, and by computation:

$$P(100) = 20756.49320, \quad S(100) = 0.00000\ 01389.$$

Substituting these values in (11), we get:

$$\sum_{x=1}^{100} \log \Gamma(x) = 15617.19560.$$

7. The Summation of $x^p \log \Gamma(x)$. The generalization of the analysis of the preceding section can be accomplished by the introduction of the Bernoulli polynomials described in Section 5 of Chapter 3.

To accomplish this we make use of the formula

$$\Delta u_x v_x = v_x \Delta u_x + u_{x+1} \Delta v_x, \tag{1}$$

which we apply to evaluating the difference of the function

$$F(x) = \frac{B_{p+1}(x)}{p+1} \log \Gamma(x). \tag{2}$$

Observing that

$$\Delta \frac{B_{p+1}(x)}{p+1} = x^p, \text{ and } \Delta \log \Gamma(x) = \log x, \tag{3}$$

we get from (1)

$$\Delta F(x) = x^p \log \Gamma(x) + \frac{B_{p+1}(x+1)}{p+1} \log x.$$

Taking sums of both sides of this equation, we obtain the desired formula

$$\sum x^p \log \Gamma(x) = \tag{4}$$
$$\frac{1}{p+1} [B_{p+1}(x) \log \Gamma(x) - \sum B_{p+1}(x+1) \log x].$$

This can be simplified somewhat further by writing

$$B_{p+1}(x+1) = B_{p+1}(x) + (p+1) x^p,$$

from which we get

$$\sum x^p \log \Gamma(x) = \tag{5}$$
$$\frac{1}{p+1} \left[B_{p+1}(x) \log \Gamma(x) - \sum B_{p+1}(x) \log x \right] - \sum x^p \log x.$$

In the application of this formula, we should observe that the constant term in $B_{p+1}(x)$ can be disregarded since $\log \Gamma(x) = \Sigma \log x$. A few special cases are given below as follows:

$$\sum \log \Gamma(x) = (x - 1) \log \Gamma(x) - \sum x \log x.$$

$$\sum x \log \Gamma(x) = \tfrac{1}{2} \left[x(x - 1) \log \Gamma(x) - \sum (x^2 + x) \log x \right].$$

$$\sum x^2 \log \Gamma(x) = \frac{1}{6} \left[x(x - 1)(2x - 1) \log \Gamma(x) \right. \tag{6}$$
$$\left. - \sum (2x^3 + 3x^2 + x) \log x \right].$$

$$\sum x^3 \log \Gamma(x) = \tfrac{1}{4} \left[x^2 (x - 1)^2 \log \Gamma(x) \right.$$
$$\left. - \sum (x^4 + 2x^3 + x^2) \log x \right].$$

$$\sum x^4 \log \Gamma(x) = \frac{1}{30} \left[x(x - 1)(2x - 1)(3x^2 - 3x - 1) \log \Gamma(x) \right.$$
$$\left. - \sum (6x^5 + 15x^4 + 10x^3 - x) \log x \right].$$

Since the problem of the summation of $x^p \log \Gamma(x)$ has now been reduced to the problem of the summation of $x^p \log x$, we shall now consider the evaluation of the sum.

$$\sum x^p \log x,$$

an example of which has already been given in Section 6.

8. *The Summation of $x^p \log x$.* The problem of the summation of $x^p \log x$ was considered very extensively in 1933 by L. Bendersky[*] who called the sums

$$L_p(n) = \sum_{x=1}^{n} x^p \log x, \quad p = 0, 1, 2, \cdots , \tag{1}$$

the logarithms of generalized gamma functions and denoted them by the symbol $\log \Gamma_p(n + 1)$. This is not a fortunate notation, however, since it conflicts with that of the imcomplete gamma function.

---

[*]"Sur la fonction gamma généralisée," *Acta Math.*, Vol. 61, 1933, pp. 263-322.

If one now substitutes $x^p \log x$ in the Euler-Maclaurin formula, he will obtain an asymptotic formula from which the following special cases are readily obtained:

$$\sum_{x=1}^{n} \log x = K_0 + (n + \tfrac{1}{2}) \log n - n$$
$$+ \frac{B_1}{1 \cdot 2} \frac{1}{n} - \frac{B_2}{3 \cdot 4} \frac{1}{n^3} + \frac{B_3}{5 \cdot 6} \frac{1}{n^5} - \cdots ,$$

$$\sum_{x=1}^{n} x \log x = K_1 + \left(\tfrac{1}{2} n^2 + \tfrac{1}{2} n + \frac{1}{12} \right) \log n - \tfrac{1}{4} n^2$$
$$+ \frac{B_2}{4!} \frac{1}{n^2} - \frac{3! \cdot B_3}{6!} \frac{1}{n^4} + \frac{5! \cdot B_4}{8!} \frac{1}{n^6} - \cdots ,$$

$$\sum_{x=1}^{n} x^2 \log x = K_2 + \left(\tfrac{1}{3} n^3 + \tfrac{1}{2} n^2 + \tfrac{1}{6} n \right) \log n - \frac{1}{9} n^3 + \frac{1}{12} n$$
$$- 2 \left(\frac{B_2}{4!} \frac{1}{n} - \frac{2! \cdot B_3}{6!} \frac{1}{n^3} + \frac{4! \cdot B_4}{8!} \frac{1}{n^5} - \cdots \right),$$

$$\sum_{x=1}^{n} x^3 \log x = K_3 + \left(\tfrac{1}{4} n^4 + \tfrac{1}{4} n^3 + \tfrac{1}{4} n^2 - \frac{1}{120} \right) \log n - \frac{1}{16} n^4$$
$$+ \frac{1}{12} n^2 - 6 \left(\frac{B_3}{6!} \frac{1}{n^2} - \frac{3! B_4}{8!} \frac{1}{n^4} + \frac{5! B_5}{10!} \frac{1}{n^6} - \cdots \right),$$

(2)

where the K_m are the following constants:

$$K_0 = 0.91893\ 85333 = \tfrac{1}{2} \log (2\pi),$$
$$K_1 = 0.24875\ 44770 = \log \text{ of Glaisher's number,}$$
$$K_2 = 0.03044\ 84571,$$
$$K_3 = -0.02065\ 63479.$$

(3)

In a manner differing in no essential manner from that already described in Section 6, the asymptotic series in each of these formulas can be represented by integrals. Using the symbol $P_p(n)$ to denote the functional part of each formula, that is,

$$P_0(n) = (n + \tfrac{1}{2}) \log n - n,$$
$$P_2(n) = \left(\tfrac{1}{2} n^2 + \tfrac{1}{2} n + \frac{1}{12} \right) \log n - \tfrac{1}{4} n^2, \text{ etc.}$$

and writing

$$\phi(s) = \frac{s}{e^s - 1},\tag{4}$$

the sums given in (2) can be expressed as follows:

$$\sum_{x=1}^{n} \log x = K_0 + P_0(n) + \int_0^\infty e^{-ns} \left[-1 + \tfrac{1}{2}s + \phi(s)\right] \frac{ds}{s^2},$$

$$\sum_{x=1}^{n} x \log x = K_1 + P_1(n)$$
$$+ \int_0^\infty e^{-ns} \left[1 - \tfrac{1}{2}s + \frac{B_1}{2!}s^2 - \phi(s)\right] \frac{ds}{s^3},$$

$$\sum_{x=1}^{n} x^2 \log x = K_2 + P_2(n)\tag{5}$$
$$+ 2\int_0^\infty e^{-ns} \left[-1 + \tfrac{1}{2}s - \frac{B_1}{2!}s^2 + \phi(s)\right] \frac{ds}{s^4},$$

$$\sum_{x=1}^{n} x^3 \log x = K_3 + P_3(n)$$
$$+ 6\int_0^\infty e^{-ns} \left[1 - \tfrac{1}{2}s + \frac{B_1}{2!}s^2 - \frac{B_2}{4!}s^4 - \phi(s)\right] \frac{ds}{s^5}.$$

CHAPTER 5

INFINITE SUMS

1. *Infinite Sums.* In preceding chapters we have been concerned for the most part with the problem of the evaluation of *finite sums*, that is to say, sums of the form

$$S_n = \sum_{x=0}^{n} u_x, \tag{1}$$

where n is finite. In a few cases, however, n was allowed to increase without limit and $S = S_\infty$ was thus the limiting value of the sum of an infinite number of terms. Such series are called *infinite series* and we shall refer to S as an *infinite sum*.

Infinite series present special problems of their own, which are concerned with the question of their *convergence* or *divergence*. If S_n approaches a finite limit S as $n \to \infty$, then we say that the series converges. Otherwise we say that the series diverges.

For example, in the case of the following three series,

$$S = 1 + \frac{1}{2^2} + \frac{1}{3^2} + \frac{1}{4^2} + \cdots + \frac{1}{n^2} + \cdots,$$

$$S' = 1 + \frac{1}{2} + \frac{1}{3} + \frac{1}{4} + \cdots + \frac{1}{n} + \cdots,$$

$$S'' = 1 - 1 + 1 - 1 + \cdots + (-1)^n + \cdots,$$

the first converges to $\pi^2/6$, and the second diverges to infinity. In the third series the *partial sums*, S_n, remain bounded, but approach no limit as $n \to \infty$. The series is thus divergent.

Although our principal concern in this chapter is to determine the limit of infinite series, that is to say, to evaluate infinite sums, after the existence of such limits has been proved, it will be convenient to describe a number of the standard tests for convergence. Proofs will not be given since they are available in many readily accessible places.

2. *Tests for Convergence of Infinite Series.* We shall assume the existence of a sequence of real numbers,

$$u_0, u_1, u_2, u_3, \cdots, u_n, \cdots \tag{1}$$

which are so constructed that u_n is uniquely defined for each value of n. From this sequence we now construct the series:

$$S = u_0 + u_1 + u_2 + \cdots + u_n + \cdots. \tag{2}$$

A few of the well known tests for the convergence (or divergence) of this series are described as follows:

(1) If the terms of the series alternate in sign, and if $u_n \to 0$ as $n \to \infty$, then the series converges.

For example, the series

$$1 - \frac{1}{2} + \frac{1}{3} - \frac{1}{4} + \frac{1}{5} - \cdots + (-1)^{n-1} \frac{1}{n} + \cdots,$$

converges, but the series

$$1 - 2 + 3 - 4 + 5 - \cdots,$$

does not converge.

(2) If after a certain value of n the terms of series (2) are all positive, and if they are term-by-term less than the corresponding terms of a known convergent series of positive terms, then series (1) converges. If, however, they are term-by-term greater than the terms of a known divergent series of positive terms, then series (2) diverges. This is know as the *comparison test*.

For example, the series

$$1 + \frac{1}{(2)^{1/2}} + \frac{1}{(3)^{1/2}} + \frac{1}{(4)^{1/2}} + \frac{1}{(5)^{1/2}} + \cdots + \frac{1}{(n)^{1/2}} + \cdots,$$

is divergent, since it is term-by-term greater than the terms of the harmonic series

$$1 + \frac{1}{2} + \frac{1}{3} + \frac{1}{4} + \frac{1}{5} + \cdots + \frac{1}{n} + \cdots,$$

which is known to be divergent.

(3) If the limit

$$\lim_{n \to \infty} \left| \frac{u_{n+1}}{u_n} \right| = R,$$

exists, then series (2) converges if $R < 1$, but diverges if $R > 1$. This is commonly called the *ratio test*.

For example, the ratio u_{n+1}/u_n for the geometric series

$$a + ar + ar^2 + ar^3 + \cdots + ar^n + \cdots$$

is the constant r. Hence the series converges when $|r| > 1$, but diverges when $|r| < 1$.

(4) If the limit

$$\lim_{n \to \infty} |u_n|^{1/n} = R,$$

exists, then the series converges if $R < 1$, but diverges if $R > 1$. This has been variously called the *radical test* and *Cauchy's test*, although the ratio test was also given by Cauchy in his famous *Algebra* published in 1821. But it appears that the ratio test had actually been published as early as 1768 by J. D'Alembert. It can be proved that when the limits defined in the ratio test and Cauchy's test both exist, they are equal.

For example, applying this test to the series

$$\frac{1 \cdot 2}{3} + \frac{2 \cdot 3}{3^2} + \frac{3 \cdot 4}{3^3} + \cdots + \frac{n(n+1)}{3^n} + \cdots ,$$

we obtain

$$\lim_{n \to \infty} \left(\frac{n(n+1)}{3^n} \right)^{1/n} = \frac{1}{3}.$$

Hence the series converges.

(5) If the ratio and Cauchy's tests fail, that is if $R = 1$, a more delicate test is provided as follow:

Let us write the absolute value of the ratio u_n / u_{n+1} in the form:

$$\left| \frac{u_n}{u_{n+1}} \right| = 1 + a_n.$$

If it happens that $\lim_{n \to \infty} a_n = 0$, that is, if $R = 1$ in either of the preceding tests, series (2) will converge if

$$\lim_{n \to \infty} n \, a_n = \rho > 1;$$

but the series will diverge if

$$\lim_{n \to \infty} n \, a_n = \rho < 1.$$

For example, we apply the test to the series

$$\frac{1}{1 \cdot 2} + \frac{1}{2 \cdot 3} + \frac{1}{3 \cdot 4} + \cdots + \frac{1}{n(n+1)} + \cdots .$$

Thus we compute

$$\left| \frac{u_n}{u_{n+1}} \right| = \frac{(n+1)(n+2)}{n(n+1)} = 1 + \frac{2}{n}.$$

Hence, since $a_n = 2/n$, we have $\rho = 2$ and the series converges.

This test is called *Raabe's test* after J. L. Raabe (1801-1859), who stated it in 1832. It is also sometimes called Duhamel's test, since it was discovered independently by J. M. C. Duhamel (1797-1872) in 1838.

(6) In case the test given in (5) fails when $\rho = 1$, the following criteria can then be applied. Let us write

$$n\, a_n = 1 + b_n,$$

where $\lim_{n\to\infty} b_n = 0$. Then the series will converge provided

$$\lim_{n\to\infty} b_n \log_e n > 1;$$

but the series will diverge provided

$$\lim_{n\to\infty} b_n \log_a n < 1.$$

By this test a series of the following type:

$$1 + \frac{1}{2\,(\log 2)^p} + \frac{1}{3\,(\log 3)^p} + \cdots + \frac{1}{n\,(\log n)^p} + \cdots ,$$

can be proved convergent if $p > 1$, and divergent if $p < 1$. The test fails for $p = 1$, but by the next test can be shown to diverge when $p = 1$.

Thus, computing $b_n \log n$, we have

$$b_n \log n = (n + 1) \log n \left[\left\{ \frac{\log\,(n + 1)}{\log n} \right\}^p - 1 \right],$$

$$= (n + 1) \log n \left[\left\{ \frac{\log n + \log\,(1 + 1/n)}{\log n} \right\}^p - 1 \right],$$

$$= (n + 1) \log n \left[\left\{ 1 + \frac{\log\,(1 + 1/n)}{\log n} \right\}^p - 1 \right].$$

When n is sufficiently large, we can replace $\log\,(1 + 1/n)$ by $1/n$ and thus obtain

$$b_n \log n \sim (n + 1) \log n \left[\left\{ 1 + \frac{1}{n \log n} \right\}^p - 1 \right],$$

$$\sim (n + 1) \log n \left[1 + \frac{p}{n \log n} - 1 \right],$$

$$\sim p\, \frac{n + 1}{n}.$$

Hence, we have the limit

$$\lim_{n\to\infty} b_n \log n = p$$

and the conclusions stated above with respect to the convergence of the series follow from the conditions of the test.

(7) A very useful test for the convergence (or divergence) of a series is provided by what is called the *integral test*. It can be stated as follows:

If, after some value of n, all the terms of the series (2) are positive and monotonically decreasing, and if the integral

$$\int_a^\infty u_n \, dn$$

exists, where a is some conveniently chosen constant, then the series converges. Otherwise, the series diverges.

As an example, let us apply this test to the case of the series discussed under (6) above. We first evaluate the following integral:

$$I(x) = \int_2^x \frac{dn}{n \, (\log n)^p},$$

$$= \frac{1}{1-p} \, [(\log x)^{1-p} - (\log 2)^{1-p}], \quad p \neq 1,$$

$$= \log \log x - \log \log 2, \quad p = 1.$$

From these equations we see that $\lim_{x \to \infty} I(x)$ exists when $p > 1$, but that $\lim_{x \to \infty} I(x) = \infty$, when $p \leqslant 1$.

(8) If the terms of the series (2) can be written in the following form:

$$u_0 = e_0 a_0, \; u_1 = e_1 a_1, \; u_2 = e_2 a_2, \cdots, \; u_n = e_n a_n, \cdots,$$

where e_0, e_1, e_2, etc. form a monotonically decreasingly set of positive numbers such that $\lim_{n \to \infty} e_n = 0$, and if the sum

$$S_n = a_0 + a_1 + a_2 + \cdots + a_n,$$

remains bounded as $n \to \infty$, then the series

$$u_0 + u_1 + u_2 + \cdots + u_n + \cdots,$$

converges.

This is called *Dirichlet's test* after P. G. L. Dirichlet (1805-1859), who first stated it. This test is especially applicable to series the terms of which are not all of the same sign. For example, let us consider the Fourier series:

$$A_1 \sin x + A_2 \sin 2x + A_3 \sin 3x + \cdots + A_n \sin nx + \cdots,$$

where the coefficients are A_1, A_2, A_3, etc. form a monotonically decreasing set of positive numbers converging to zero. They might be, for example, 1, 1/2, 1/3, etc.

We now compute

$$S_n = \sin x + \sin 2x + \sin 3x + \cdots + \sin nx,$$
$$= \left[\sin \tfrac{1}{2}nx \sin \tfrac{1}{2}(n+1) x\right] / \sin \tfrac{1}{2}x.$$

Since this sum remains bounded for all values of x as $n \to \infty$, it follows from the conditions of the test that the Fourier series converges.

(9) Another test, called *Abel's test* after N. H. Abel (1802-1829), is stated as follows: Using the notation of the preceding test, we assume that S_n approaches a limit S as $n \to \infty$. In this case, if the quantities e_1, e_2, e_3, etc. form a monotonically increasing or decreasing series bounded by a value E independent of n, then the series

$$u_0 + u_1 + u_2 + \cdots + u_n + \cdots$$

converges.

(10) One often encounters a function defined as follows:

$$S(x) = U_1(x) + U_2(x) + \cdots + U_n(x) + \cdots , \qquad (3)$$

where the individual terms are functions defined within an interval: $a \leqslant x \leqslant b$.

Such a series is said to be *uniformly convergent* if, for any arbitrarily chosen positive number ϵ, there exists an N such that

$$|S(x) - S_n(x)| < \epsilon, \qquad (4)$$

for every $n > N$ and every x in (a, b).

A test for uniform convergence, called the *Weierstrass test*, is the following:

If there exists a convergent series of positive terms,

$$M_1 + M_2 + \cdots + M_n + \cdots , \qquad (5)$$

such that $|U_n(x)| \leqslant M_n$ for all values of x in (a, b), then (3) converges uniformly in the interval.

Series (5) is called a *majorante*. It can be readily proved that if the individual terms are continuous functions of x, then $S(x)$ is a continuous function in (a, b) if the series converges uniformly.

For example, the series

$$\frac{\sin x}{1^2} - \frac{\sin 3x}{3^2} + \frac{\sin 5x}{5^2} - \frac{\sin 7x}{7^2} + \cdots$$

converges uniformly in the interval $-\pi \leqslant x \leqslant \pi$ and defines a continuous function, since the series

$$1 + \frac{1}{3^2} + \frac{1}{5^2} + \frac{1}{7^2} + \cdots$$

forms a majorante for it.

But the following series, which converges throughout the interval $-\pi \leqslant x \leqslant \pi$, does not converge uniformly:

$$\frac{\sin x}{1} + \frac{\sin 3x}{3} + \frac{\sin 5x}{5} + \frac{\sin 7x}{7} + \cdots$$

Although the terms of the series are continuous, the function which it defines is $\frac{1}{4}\pi$ in the positive interval and $-\frac{1}{4}\pi$ in the negative interval with discontinuities at $x = 0, +\pi$, and $-\pi$.

PROBLEMS

Establish the convergence or divergence of the following series:

1. $\dfrac{1}{\log 2} - \dfrac{1}{\log 3} + \dfrac{1}{\log 4} - \dfrac{1}{\log 5} + \cdots$

2. $1 - \dfrac{3}{4} + \dfrac{5}{8} - \dfrac{7}{12} + \dfrac{9}{16} - \cdots$

3. $\dfrac{1}{\log \log 2} - \dfrac{1}{\log \log 3} + \dfrac{1}{\log \log 4} - \dfrac{1}{\log \log 5} + \cdots$

4. $\dfrac{1 + e^{-1}}{5} - \dfrac{2 + e^{-2}}{10} + \dfrac{3 + e^{-3}}{15} - \dfrac{4 + e^{-4}}{20} + \cdots$

5. $1 + \dfrac{1}{2 \, (2)^{1/2}} + \dfrac{1}{3(3)^{1/2}} + \dfrac{1}{4(4)^{1/2}} + \cdots$

6. $1 + \dfrac{1}{2^2 - 2} + \dfrac{1}{3^2 - 3} + \dfrac{1}{4^2 - 4} + \cdots$

7. $\displaystyle\sum_{n=2}^{\infty} \dfrac{(n)^{1/2}}{n^2 - 1}.$ 8. $\displaystyle\sum_{n=1}^{\infty} \sin\left(\dfrac{1}{n^2}\right).$

9. $\dfrac{\log 2}{1 \cdot 2} + \dfrac{\log 3}{2 \cdot 3} + \dfrac{\log 4}{3 \cdot 4} + \cdots + \dfrac{\log (n + 1)}{n(n + 1)} + \cdots$

10. $\dfrac{1^2}{2^p} + \dfrac{2^2}{3^p} + \dfrac{3^2}{4^p} + \dfrac{4^2}{5^p} + \cdots + \dfrac{n^2}{(n + 1)^p} + \cdots$

11. $\dfrac{1}{3^2 - 2^2} + \dfrac{1}{3^3 - 2^3} + \dfrac{1}{3^4 - 2^4} + \cdots + \dfrac{1}{3^n - 2^n} + \cdots$

12. $\dfrac{1 \cdot 3}{2 \cdot 4} + \dfrac{1 \cdot 3 \cdot 5}{2 \cdot 4 \cdot 6} + \dfrac{1 \cdot 3 \cdot 5 \cdot 7}{2 \cdot 4 \cdot 6 \cdot 8} + \cdots$

13. $\dfrac{1}{2} + \dfrac{(2)^{1/2}}{3} + \dfrac{(3)^{1/2}}{10} + \dfrac{(4)^{1/2}}{17} + \cdots + \dfrac{(n)^{1/2}}{n^2 + 1} + \cdots$

14. $\dfrac{1}{3 \cdot 5} + \dfrac{2}{5 \cdot 7} + \dfrac{3}{7 \cdot 9} + \cdots + \dfrac{n}{(2n + 1)(2n + 3)} + \cdots$

Determine the values of x for which the following series converge:

15. $\dfrac{2x}{2} + \dfrac{(2x)^2}{3} + \dfrac{(2x)^3}{4} + \cdots + \dfrac{(2x)^n}{n + 1} + \cdots$

16. $1 + \dfrac{(x + 2)}{1^2 2} + \dfrac{(x + 2)^2}{2^2 2^2} + \dfrac{(x + 2)^3}{3^2 2^3} + \dfrac{(x + 2)^4}{4^2 2^4} + \cdots$

17. $1 + x + \dfrac{2^2}{2!} x^2 + \dfrac{3^3}{3!} x^3 + \cdots + \dfrac{n^n}{n!} x^n + \cdots$

18. $\frac{1}{2}x - \dfrac{1}{1!2!} (\frac{1}{2}x)^3 + \dfrac{1}{2!3!} (\frac{1}{2}x)^5 - \dfrac{1}{3!4!} (\frac{1}{2}x)^7 + \cdots$

19. Discuss the convergence of the following series:

$$F(x) = \frac{\cos 2x}{1 \cdot 3} + \frac{\cos 4x}{3 \cdot 5} + \frac{\cos 6x}{5 \cdot 7} + \cdots , 0 < x < \pi.$$

Show that its first derivative, but not its second, converges. Given that $F(x) = \frac{1}{2} - \frac{1}{4}\pi \sin x$, discuss the behavior of the derivative series at $x = 0$.

3. *The Summation of Infinite Series.* We shall consider in the following pages a few of the methods available for the evaluation of

$$S = \lim_{n \to \infty} S_n, \tag{1}$$

where we define

$$S_n = \sum_{x=0}^{n} u(x). \tag{2}$$

We shall assume that $u(x)$ satisfies one of the tests of convergence given in Section 2 and shall limit our attention to the problem of evaluating S.

Many ingenious special methods applicable to the summation of particular functions are found in the literature of the subject. Such, for example, are those which have been adapted to the proof of the astonishing formulas of S. Ramanujan (1887-1920), typical of which are the following:[*]

$$\sum_{n=1}^{\infty} \frac{n^{13}}{e^{2n\pi} - 1} = \frac{1}{24}, \quad \sum_{n=1}^{\infty} \frac{\coth n\pi}{n^7} = \frac{19 \pi^7}{56700}.$$

[*]See *The Collected Papers of Srinivasas Ramanujan.* Cambridge, 1927.

But general methods, adaptable to a wide class of sums, are more useful and a few of these we shall consider in the next sections. These we list below as follows:

(A) *The Method of Taylor's Series.* This method consists in comparing a given function $u(n)$ with the nth term of the Taylor's expansion of $f(x)$ and identifying the particular function which provides the limit:

$$\lim_{x \to 0} D^n f(x) = n ! \, u(n).$$

(B) *The Method of Inverse Differences.* This method depends essentially upon finding a function $U(x)$, which satisfies the difference equation:

$$\Delta U(x) = u(x).$$

(C) *The Summation of Powers of Reciprocal Roots.* By certain properties of entire functions, it is possible to find the sums of the powers of the reciprocal roots of the equation $f(x) = 0$, where $f(x)$ is an entire function.

(D) *Summation by Means of Poisson's Formula and the Euler-Maclaurin Expansion.* These powerful devices are adapted to certain classical sums and a comparison is made of their resemblances and differences.

4. *The Method of Taylor's Series.* The greatest generator of infinite sums has been Taylor's series, which has already been described in Section 7 of Chapter 1. Ever since its discovery in 1712, there has been continuous application of it to the expansion of special functions. This has resulted in an impressive collection of infinite sums.

Since many excellent works are available describing the theory and application of Taylor's series, we shall give here only a very superficial account of the subject. For our purpose, let us write

$$f(x) = \sum_{n=0}^{\infty} u(n) \, x^n, \tag{1}$$

which is a Taylor's series if $f(x)$ is an infinitely differentiable function and $u(n)$ is defined as the following limit:

$$u(n) = \lim_{x \to 0} \frac{D^n f(x)}{n !}. * \tag{2}$$

---

*This special case of Taylor's theorem is usually referred to as *Maclaurin's series.*

This series converges uniformly within a circular area in the complex plane, the center of which is the origin and the radius the distance from the origin to the nearest singular point of the function. This fact is often very useful in determing the radius of convergence of series (1).

For example, it is a matter of some difficulty, involving a knowledge of the properties of the Bernoulli numbers, B_n, to determine by direct computation the radius of convergence of the following series:

$$\frac{x}{e^x - 1} = 1 - \tfrac{1}{2}x + \sum_{n=1}^{\infty} (-1)^{n+1} \frac{B_n}{(2n)!} x^{2n}. \tag{3}$$

But when we see that the denominator of the function vanishes for $x = 2\pi i$ and that there is no other singularity nearer to the origin than this, we conclude that the series converges for all values of x such that $|x| < 2\pi$.

While it is a comparatively simple matter to generate a Taylor's series when $f(x)$ is given, it is quite another problem to find $f(x)$ when $u(n)$ is given. The nature of the difficulty can be inferred from the following analysis:

If we introduce the complex variable $x = r e^{\theta i} = r (\cos \theta + i \sin \theta)$ into equation (1), the left hand member can be written

$$f(x) = u(r, \theta) + i v(r, \theta),$$

and the right hand member reduces to the following sums:

$$\sum_{n=0}^{\infty} u(n) r^n \cos n\theta + i \sum_{n=0}^{\infty} u(n) r^n \sin n\theta.$$

Equating real and imaginary parts, we obtain the two equations:

$$\sum_{n=0}^{\infty} u(n) r^n \cos n\theta = u(r, \theta), \tag{4}$$

$$\sum_{n=0}^{\infty} u(n) r^n \sin n\theta = v(r, \theta).$$

If each of these is multiplied respectively by $\cos n\theta$ and $\sin n\theta$ and integrated between $-\pi$ and $+\pi$, the following equations are obtained:

$$\int_{-\pi}^{\pi} u(r, \theta) \cos n\theta \, d\theta = \pi \, u(n) \, r^n,$$

$$\int_{-\pi}^{\pi} v(r, \theta) \sin n\theta \, d\theta = \pi \, u(n) \, r^n.$$

If now these equations could be solved for u and v, then $f(x)$ could be constructed from them. But unfortunately they are Fredholm integral equations of first kind with kernels respectively equal to $\cos n\theta$ and $\sin n\theta$ and such equations, except in special cases, have proved to be intractable.

However, when $f(x)$ is known equations (4) are very useful in obtaining the values of Fourier series. This is illustrated by the following examples.

Example 1. If $u(0) = 0$, $u(n) = 1/n$, $n > 1$, then $f(x) = -\log(1 - x)$. Hence we compute

$$-\log(1 - r\,e^{\theta i}) = -\log(1 - r\cos\theta - i\,r\sin\theta),$$

$$= -\tfrac{1}{2}\log(1 - 2r\cos\theta + r^2) + i\arctan\left(\frac{r\sin\theta}{1 - r\cos\theta}\right).$$

If $r = 1$, we then obtain

$$u(1, 0) = -\log(2\sin\tfrac{1}{2}\theta), \quad v(1, \theta) = \arctan\left(\frac{\sin\theta}{1 - \cos\theta}\right) = \tfrac{1}{2}\pi - \tfrac{1}{2}\theta,$$

and from these, referring to (4), we get

$$\sum_{n=1}^{\infty} \frac{\cos n\theta}{n} = -\log 2 - \log\sin\tfrac{1}{2}\theta, \quad \sum_{n=1}^{\infty} \frac{\sin n\theta}{n} = \tfrac{1}{2}\pi - \tfrac{1}{2}\theta.$$

Example 2. Given the series:

$$C(\theta) = 1 - \frac{1}{2}\cos\theta + \frac{1}{2^2}\cos 2\theta - \frac{1}{2^3}\cos 3\theta + \cdots,$$

show that $C(\theta) = (4 + 2\cos\theta)/(5 + 4\cos\theta)$.

Solution: From the coefficients of the given series we construct the following function:

$$f(x) = 1 - \frac{1}{2}x + \frac{1}{2^2}x^2 - \frac{1}{2^3}x^3 + \cdots = \frac{2}{2 + x}.$$

Replacing x by $\cos\theta + i\sin\theta$, we get

$$f(x) = \frac{4 + 2\cos\theta - i\sin\theta}{5 + 4\cos\theta}.$$

The real part of this expression is $C(\theta)$.

PROBLEMS

1. Show that

$$\sum_{n=1}^{\infty} \frac{x^n}{n^2} = -\int_0^1 \frac{\log(1 - xp)}{p}\,dp.$$

2. Prove the following:

$$\sum_{n=1}^{\infty} \frac{1^3 + 2^3 + \cdots + n^3}{n!}\,x^n = e^x\left(x + \frac{7}{2}x^2 + 2x^3 + \frac{1}{4}x^4\right).$$

3. Prove that

$$D^n \sin px = p^n \sin (px + \tfrac{1}{2}n\pi), \quad D^n \cos px = p^n \cos (px + \tfrac{1}{2}n\pi),$$

and from these obtain the expansions of $\sin px$ and $\cos px$.

4. Show that

$$D^n e^{ax} \sin mx = (a^2 + m^2)^{\frac{1}{2}n} e^{ax} \sin (mx + n\phi),$$

where $\tan \phi = m/a$. From this obtain the expansion of $e^{ax} \sin ax$.

5. Prove the following:

$$D^n \tan^{-1} \frac{x}{a} = (-1)^{n-1} \frac{\sin^n \phi \sin n\phi}{a^n},$$

where $\cot \phi = x/a$. From this obtain the expansion of $\tan^{-1} x/a$.

6. From the expansion about the origin of $(1 + x)^{\frac{1}{2}}$, establish the following:

$$1 + \frac{1}{2} \cos \theta - \frac{1}{2 \cdot 4} \cos 2\theta + \frac{1 \cdot 3}{2 \cdot 4 \cdot 6} \cos 3\theta + \cdots = (2 \cos \tfrac{1}{2}\theta)^{1/2} \sin \tfrac{1}{4}\theta.$$

7. Prove the following:

$$u(x, y) = \frac{4}{\pi} \left[\sin \frac{\pi x}{a} e^{-\pi y/a} + \frac{1}{3} \sin \frac{3\pi x}{a} e^{-3\pi y/a} + \frac{1}{5} \sin \frac{5\pi x}{a} e^{-5\pi y/a} + \cdots \right],$$

$$= \frac{2}{\pi} \arctan \left[\frac{\sin (\pi x/a)}{\sinh (\pi y/a)} \right]. \quad Hint: \quad \text{Consider the expansion of}$$

$f(x) = \log [(1 + x)/(1 - x)]$.

5. *The Method of Inverse Differences.* If $u(x)$ is an arbitrarily given function of x, which is continuous over the infinite real interval: $x^2 \geqslant 0$, and if $U(x)$ is any continuous solution of the equation

$$\Delta U(x) = u(x), \tag{1}$$

then we have

$$S(a, b) = \sum_{x=a}^{b} u(x) = U(a) - U(b). \tag{2}$$

If the following limits exist:

$$\lim_{a \to -\infty} U(a) = U_1, \quad \lim_{b \to \infty} U(b) = U_2, \tag{3}$$

then the infinite sum S becomes

$$S = \sum_{x=-\infty}^{\infty} u(x) = U_1 - U_2. \tag{4}$$

Examples of such sums have already been given in Sections 6 and 8 of Chapter 2 and in Section 5 of Chapter 4. Others illustrating this method are given below as follows:

Example 1. Show that

$$S_p = \sum_{x=0}^{\infty} a^x \, x^{(p)} = \frac{p! \, a^p}{(1-a)^{p+1}}, \quad 0 < a < 1, \, p \geqslant 0. \tag{5}$$

This sum, it will be observed, is the analogue of the well known integral

$$\int_0^{\infty} e^{-ax} x^p \, dx = \frac{p!}{a^{p+1}}.$$

Solution: Taking note of the formula

$$\sum a^x \, x^{(p)} = x^{(p)} \frac{a^x}{a-1} - \frac{pa}{a-1} \sum a^x \, x^{(p-1)}, \quad a \neq 1, \tag{6}$$

and observing that

$$\lim_{x \to \infty} a^x \, x^{(p)} = 0, \tag{7}$$

when a is a positive number less than 1 and p is zero or any positive integer, we have

$$S_p = \frac{ap}{1-a} S_{p-1}. \tag{8}$$

By direct calculation we have

$$S_0 = \sum_{x=0}^{\infty} a^x = 1/(1-a), \tag{9}$$

and formula (5) is then readily established by indiction.

Example 2. The results given in Example 1 can be extended if $x^{(p)}$ is replaced by its generalized form [See (1), Section 3, Chapter 2] as follows:

$$x^{(p)} = \frac{\Gamma(x+1)}{\Gamma(x-p+1)}. \tag{10}$$

The following formula is readily established by differencing:

$$\sum a^x \frac{\Gamma(x+1)}{\Gamma(x-p+1)} = \frac{pa}{1-a} \sum a^x \frac{\Gamma(x+1)}{\Gamma(x-p+2)} - \frac{a^x}{1-a} \frac{\Gamma(x+1)}{\Gamma(x-p+1)}. \tag{11}$$

Summing between 0 and ∞, and taking account of (7), we obtain

$$S_p = \frac{pa}{1-a} S_{p-1} + \frac{1}{1-a} \frac{1}{\Gamma(1-p)}, \quad 0 < a < 1, \, p \text{ arbitrary.} \tag{12}$$

where we abbreviate

$$S_p = \sum_{x=0}^{\infty} a^x \frac{\Gamma(x+1)}{\Gamma(x-p+1)}. \tag{13}$$

If p is any positive integer, the last term in (12) is zero and we obtain formula (8).
If p is a negative integer, let us say $p = -n$, and if we write

$$S_{-n} = \sum_{x=0}^{\infty} a^x \frac{\Gamma(x+1)}{\Gamma(x+n+1)}. \tag{14}$$

then (12) is replaced by the following:

$$S_{-n-1} = \frac{(a-1)}{na} S_{-n} + \frac{1}{an \cdot n!}. \tag{15}$$

By direct computation

$$S_{-1} = -\frac{1}{a} \log (1-a), \tag{16}$$

whence we get from (15),

$$S_{-2} = \frac{1-a}{a^2} \log (1-a) + \frac{1}{a},$$

$$S_{-3} = -\frac{(1-a)^2}{2a^3} \log (1-a) - \frac{1-a}{2a^2} + \frac{1}{4a}, \text{ etc.}$$

Another interesting sum is obtained from (10) if we set $p = \frac{1}{2}$ and observe that

$$\Gamma(x + \tfrac{1}{2}) = \frac{2(\pi)^{1/2} \Gamma(2x)}{2^{2x} \Gamma(x)}. \tag{17}$$

We thus obtain

$$S_{\frac{1}{2}} = 1 + \frac{1}{2!}z + \frac{(2!)^2}{4!}z^2 + \frac{(3!)^2}{6!}z^3 + \cdots, \tag{18}$$

where $z = 4a$.

Example 3. Evaluate the sum

$$S = \frac{1}{2} \tan (\theta/2) + \frac{1}{2^2} \tan (\theta/2^2) + \frac{1}{2^3} \tan (\theta/2^3) + \cdots, \quad |\theta| < \pi. \tag{19}$$

Solution: From the *Table of Finite Sums* (E, 17), we have

$$\sum \frac{\tan \theta/2^{x+1}}{2^{x+1}} = \frac{1}{2^x \tan \theta/2^x}.$$

If $|\theta| < \pi$, then every term in the series is finite and the convergence of the series is readily established. Observing finally that

$$\lim_{x \to \infty} \frac{1}{2^x \tan \theta/2^x} = \frac{1}{\theta},$$

we thus have

$$S = \frac{1}{\theta} - \frac{1}{\tan \theta}. \tag{20}$$

PROBLEMS

1. Show that

$$1 + \frac{1}{2} + \frac{2}{2^2} + \frac{3}{2^3} + \cdots = 3.$$

2. Prove that

$$\sum_{x=0}^{\infty} a^x \cos \theta x = \frac{1 - a \cos \theta}{a^2 - 2a \cos \theta + 1}, \quad |a| < 1. \tag{21}$$

3. Evaluate the following sum:

$$\sum_{x=0}^{\infty} \frac{a^x}{(x+2)(x+3)}, \quad |a| < 1.$$

$$Ans. \ -\frac{1}{2a} + \frac{1}{a^2} + \frac{1-a}{a^3} \log(1-a).$$

4. Determine the range of convergence of series (18).

6. *The Sums of Powers of Reciprocal Roots.* In Chapter 2, by means of the properties of the polygamma functions, we have shown how to evaluate such series as

$$\sum_{x=1}^{\infty} \frac{1}{x^{2n}} \text{ and } \sum_{x=1}^{\infty} \frac{1}{(2x-1)^{2n}}. \tag{1}$$

We shall now consider such sums from another point of view.

For this purpose let us introduce $f(x)$ as an entire function of genus p. By thus we mean that $f(x)$ has the following properties:

(1) $f(x)$ has no singularities in the finite plane and can be expanded in a Taylor's series with an infinite radius of convergence.

(2) If $f(x)$ vanishes for a set of non-zero values: $x_1, x_2, x_3 \ldots$, then the series

$$S = \sum_{x=1}^{\infty} 1/|x_n|^{k+1}, \tag{2}$$

will converge when k is any number greater than or equal to p.

(3) If k is the smallest integer for which (2) converges, then $f(x)$ can be written as follows:

$$f(x) = e^{Q(x)} \prod_{n=1}^{\infty} (1 - x/x_n) e^{x/x_n + \frac{1}{2}x^2/x_n^2 + \cdots + x^k/kx_n^k}, \tag{3}$$

where $Q(x)$ is a polynomial of degree q. The genus p of $f(x)$ is the larger of the two integers q and k.

Thus, the functions e,^{x} sin x, and cos x are entire functions of genus 1, but e^{z^2} is of genus 2. Since the reciprocal of $\Gamma(x)$ can be written as follows:

$$\frac{1}{\Gamma(x)} = x \, e^{\gamma x} \prod_{n-1}^{\infty} \left(1 + \frac{x}{n}\right) e^{-x/n}, \tag{4}$$

we see that $1/\Gamma(x)$ is an entire function of genus 1.

A theorem which we shall need in another section is the following:

Theorem 1. Let $f(x)$ be an entire function of genus p and let us define the sequence:

$$L_n = \left| f^{(n)}(x) \right|^{1/n}. \tag{5}$$

If $O[\phi(n)]$ is the order of L_n, then $\phi(n) = n^{(p-1)2p}$. *

A second theorem which is of immediate interest to us is the following:

Theorem 2. If $f(x)$ is an entire function of genus p, and if x_1, x_2, x_3, $\cdots$ form a set of zeros of $f(x)$, then the following indenity holds:

$$D^m \left\{ \frac{f'(x)}{f(x)} \right\} = - m! \sum_{n-1}^{\infty} \frac{1}{(x_n - x)^{m+1}}, \tag{6}$$

where the right hand member converges for all values of $m > p$. †

From this theorem we can now obtain the sums of the powers of the reciprocal roots of $f(x) = 0$. Thus, if we write

$$S_m = \sum_{n-1}^{\infty} \frac{1}{x_n^m}, \tag{7}$$

we obtain as an immediate consequence of the theorem

$$S_{m+1} = - \frac{1}{m!} \lim_{x \to 0} D^m \frac{f'(x)}{f(x)}, \quad m > p. \tag{8}$$

---

*By this we mean that the ratio $L_n/\phi(n)$ is bounded. For a proof of this theorem see E. Borel: *Leçons sur les fonctions entière*. Paris, 1921, Chapter 3.

† See L. V. Ahlfors: *Complex Analysis*. New York, 1953, p. 188. The case when $m = 1$, is given by Borel (*loc. cit.*), p. 32.

Example 1. Let $f(x) = \cos x$. We observe that $\cos x$ is an entire function of genus 1 and that its zeros are $\pm (2n - 1) \pi/2$. Since we have

$$\frac{f'(x)}{f(x)} = -\tan x = \qquad (9)$$

$$-\left[x + \frac{x^3}{3} + \frac{2}{15} x^5 + \cdots + \frac{2^{2m}(2^{2m} - 1) B_m}{(2m)!} x^{2m-1} + \cdots \right],$$

we obtain the following sums:

$$S_2 = \frac{8}{\pi^2} \sum_{n=1}^{\infty} \frac{1}{(2n - 1)^2} = \lim_{x \to 0} D \tan x = 1,$$

$$S_4 = \frac{32}{\pi^4} \sum_{n=1}^{\infty} \frac{1}{(2n - 1)^4} = \frac{1}{3}, \qquad (10)$$

$$S_{2m} = \frac{2^{2m+1}}{\pi^{2m}} \sum_{n=1}^{\infty} \frac{1}{(2n - 1)^{2m}} = \frac{2^{2m}(2^{2m} - 1) B_m}{(2m)!}.$$

Example 2. Let $f(x) = \sin x/x$, whence $f(x)$ is an entire function of genus 1 and its zeros are $\pm n\pi$. Since we have

$$\frac{f'(x)}{f(x)} = \cot x - \frac{1}{x} = -\left[\frac{x}{3} + \frac{x^3}{45} + \frac{2}{945} x^5 + \cdots + \frac{2^{2m} B_m}{(2m)!} x^{2m-1} + \cdots \right], \qquad (11)$$

we readily obtain

$$S_2 = \frac{2}{\pi^2} \sum_{n=1}^{\infty} \frac{1}{n^2} = \lim_{x \to 0} D \left(\frac{1}{x} - \cot x \right) = \frac{1}{3},$$

$$S_4 = \frac{2}{\pi^4} \sum_{n=1}^{\infty} \frac{1}{n^4} = \frac{1}{45}, \qquad (12)$$

$$S_{2m} = \frac{2}{\pi^{2m}} \sum_{n=1}^{\infty} \frac{1}{n^{2m}} = \frac{2^{2m} B_m}{(2m)!}.$$

In the examples just given, the application of formula (8) has been relatively simple, but this is not generally the case. Thus, the problem of evaluating S_m for the roots of $J_n(x) = 0$, where $J_n(x)$ is the Bessel function or order n, becomes very arduous for large values of n.

A considerably simpler method is provided through a consideration of the problem of evaluating the sums of the powers of roots of the following algebraic equation:

$$x^n - A x^{n-1} + B x^{n-2} - C x^{n-3} + \cdots \qquad (13)$$

$$+ (-1)^{n-1} M x + (-1)^n N = 0.$$

If we denote by S_k the following sum:

$$S_k = \sum_{i=1}^{n} x_i^k, \qquad (14)$$

where $x_1, x_2, x_3, \ldots$ are the roots of (13), then these sums are obtained by the well known *formulas of Newton*.[*]

$$S_1 - A = 0,$$

$$S_2 - A\,S_1 + 2B = 0,$$

$$S_3 - A\,S_2 + B\,S_1 - 3C = 0,$$

$$S_{n-1} - A\,S_{n-2} + B\,S_{n-3} - \cdots + (-1)^{n-1}(n-1)M = 0,$$

$$S_k - A\,S_{k-1} + B\,S_{k-2} - \cdots + (-1)^n N\,S_{k-n} \qquad (15)$$

$$= 0,\ k = n,\, n+1, \cdots$$

Solving these equations for S_k, $k = 1$ to 9, we obtain the following explicit formulas:

$$S_1 = A,$$

$$S_2 = A^2 - 2B,$$

$$S_3 = A^3 - 3AB + 3C,$$

$$S_4 = A^4 - 4A^2 B + 4AC + 2B^2 - 4D,$$

$$S_5 = A^5 - 5A^3 B + 5A^2 C + 5AB^2 - 5AD - 5BC + 5E,$$

$$S_6 = A^6 - 6A^4 B + 6A^3 C + 9A^2 B^2 - 6A^2 D - 12ABC - 2B^3$$

$$+ 6BD + 3C^2 + 6AE - 6F,$$

$$S_7 = A^7 - 7A^5 B + 7A^4 C + 14A^3 B^2 - 7A^3 D - 21A^2 BC - 7AB^3$$

$$+ 14ABD + 7AC^3 + 7A^2 E - 7AF$$

$$+ 7B^2 C - 7BE - 7CD + 7G,$$

---

[*]See M. Bôcher: *Introduction to Higher Algebra.* New York, 1921, pp. 243-244.

$$S_8 = A^8 - 8A^6B + 8A^5C + 20A^4B^2 - 8A^4D - 32A^3BC$$

$$- 16A^2B^3 + 24A^2BD + 12A^2C^2 + 8A^3E - 8A^2F$$

$$+ 24AB^2C - 16ABE - 16ACD + 8AG + 2B^4$$

$$- 8B^2D - 8BC^2 + 8BF + 8CE + 4D^2 - 8H,$$

$$S_9 = A^9 - 9A^7B + 9A^6C + 27A^5B^2 - 9A^5D - 45A^4BC$$

$$- 30A^3B^3 + 36A^3BD + 18A^3C^2 + 9A^4E - 9A^3F$$

$$+ 54A^2B^2C - 27A^2BE - 27A^2CD + 9A^2G$$

$$+ 9AB^4 - 27AB^2D - 27ABC^2 + 18ABF$$

$$+ 18ACE + 9AD^2 - 9AH - 9B^3C + 9B^2E$$

$$+ 18BCD - 9BG + 3C^3 - 9CF - 9DE + 9I.$$

If now in equation (13) x is replaced by $1/x$, the equation assumes the following form:

$$1 - Ax + Bx^2 - Cx^3 + \cdots = 0, \tag{16}$$

and S_k defined by (14) is now of the reciprocal roots of (16), that is,

$$S_k = \sum_{i=1}^{n} \frac{1}{x_i^k}. \tag{17}$$

It is now a simple generalization to replace the left hand member of (16) by an entire function of genus p and apply Theorem 2 to assure the existence of S_k, which is now evaluated by the formulas just given. This method is illustrated by the following examples:

Example 3. To find S_2 and S_4 for the function $\sin x/x$, we first observe from the expansion:

$$\frac{\sin x}{x} = 1 - \frac{1}{3!}x^2 + \frac{1}{5!}x^4 - \cdots , \tag{18}$$

that $A = C = 0$, $B = -1/3!$, $D = 1/5!$. Hence, applying the formulas, we get as in Example 2:

$$S_2 = \frac{2}{\pi^2}\sum_{n=1}^{\infty}\frac{1}{n^2} = \frac{2}{3!} = \frac{1}{3}, \quad S_4 = \frac{2}{\pi^4}\sum_{n=1}^{\infty}\frac{1}{n^4} = \frac{3}{36} - \frac{4}{120} = \frac{1}{45}.$$

Example 4. Given the following expansion:

$$\frac{2^n n!}{x^n} J_n(x) = 1 - \frac{1}{n+1} \left(\frac{x}{2}\right)^2 + \frac{1}{2!} \frac{1}{(n+1)(n+2)} \left(\frac{x}{2}\right)^4$$

$$- \frac{1}{3!} \frac{1}{(n+1)(n+2)(n+3)} \left(\frac{x}{2}\right)^6 + \cdots,$$

compute S_2, S_4, and S_6, where these denote respectively the sums of the second, fourth, and sixth powers of the reciprocals of the positive roots of $J_n(x) = 0$.

Solution: From the expansion one obtains the following values:

$$A = C = E = 0,$$

$$B = -\frac{1}{4(n+1)},$$

$$D = \frac{1}{32(n+1)(n+2)},$$

$$F = -\frac{1}{384(n+1)(n+2)(n+3)}.$$

When these quantities are substituted in the above formulas and account taken of the fact that we are computing the sums of the reciprocals of the positive roots, we obtain the following values:

$$S_2 = \frac{1}{2^2 (n+1)},$$

$$S_4 = \frac{1}{2^4 (n+1)^2 (n+2)},$$

$$S_6 = \frac{2}{2^6 (n+1)^3 (n+2)(n+3)}.$$

It is a matter of some interest to observe that the problem of finding these sums for $J_n(x)$ originates with Euler. G. N. Watson in his *Theory of Bessel Function* (1922) gives the values of S_{2m} to $m = 5$ and also the value of S_{16}, the latter being attributed to A. Cayley.[*] R. E. Shafer at the Lawrence Radiation Laboratory extended the computations to include S_{12}, S_{14}, and S_{18}, which are given as follows:

$$S_{12} = \frac{21n^3 + 181n^2 + 513n + 473}{2^{11}(n+1)^6 (n+2)^3 (n+3)^2 (n+4)(n+5)(n+6)},$$

$$S_{14} = [33n^3 + 329n^2 + 1081n + 1145]/[2^{12}(n+1)^7 (n+2)^3$$

$$(n+3)^2 (n+4)(n+5)(n+6)(n+7)],$$

[*]For these values, see Watson: p. 502.

$$S_{18} = [715n^6 + 16567n^5 + 158568n^4 + 798074n^3 + 2217079n^2$$

$$+ 3212847n + 1893046]/[2^{17} (n + 1)^9 (n + 2)^4$$

$$(n + 3)^3 (n + 4)^2 (n + 5) (n + 6)$$

$$(n + 7) (n + 8) (n + 9)].$$

PROBLEMS

1. The roots of the following equation are the squares of the roots of equation (13):

$$y^n - (A^2 - 2B)y^{n-1} + (B^2 - 2AC + 2D)y^{n-2}$$

$$- (C^3 - 2BD + 2AE - 2F)y^{n-3} \tag{19}$$

$$+ (D^2 - 2CE + 2BF - 2AG + 2H)y^{n-4} - \cdots = 0.$$

Replace the equation $\cos x = 0$ by one in which the reciprocal roots have been squared and use the coefficients of the new equation to compute S_4.

2. Apply the method of Problem 1 a second time to find the equation in which the reciprocal roots of $\cos x = 0$ have been raised to the fourth power. Use the coefficients of the new equation to compute S_8.

3. The reciprocal Gamma function has the following expansion:

$$1/\Gamma(x) = x + C_2 x^2 + C_3 x^3 + C_4 x^4 + \cdots .$$

Show that $C_2 = \gamma$, and that $C_3 = \frac{1}{2}\gamma^2 - \pi^2/12$.

4. Hermite polynomials of even degree, $h_{2n}(x)$, are defined by the equation:

$h_{2n}(x) = f^{(2n)}(x)/f(x)$, where $f(x) = e^{-\frac{1}{2}x^2}$. Explicitly they are found to be

$$h_{2n}(x) = x^{2n} - \frac{1}{2} (2n) (2n - 1) x^{2n-2}$$

$$+ \frac{1}{2 \cdot 4} (2n) (2n - 1) (2n - 2) (2n - 3) x^{2n-4} - \cdots$$

If S_k is the sum of the kth powers of the reciprocal roots, show that

$$S_2 = n, \quad S_4 = \frac{1}{3} n(2n + 1), \quad S_6 = \frac{1}{15} n(4n + 1) (2n + 1).$$

7. *Poisson's Formula.* In previous sections of this work we have indicated the usefulness of the Euler-Maclaurin formula in the processes of summation. We shall now introduce another series, which has much in common with that of Euler-Maclaurin, and in some applications is superior to it. This is *Poisson's formula*, which, for some strange reason, has been largely neglected by text book writers in spite of its obvious power in the summation of series.

Although the Euler-Maclaurin formula appears in many places, the Poisson formula is discussed in relatively few standard works. It seems to have been stated first by S. D. Poisson in 1827*, although it may have been known earlier to K. F. Gauss.† When it was first observed by A. L. Cauchy, his admiration was very great. He said that it was a discovery worthy of the genius of Laplace. But strange to say, in spite of this, the number of analysts, who have been attracted to the theorem, and who have used it, has been suprisingly small.§

For our purpose *Poisson's formula* can be stated formally as follows:

If $f(x)$ and $F(x)$ are cosine transforms of one another, that is if

$$f(x) = \left(\frac{2}{\pi}\right)^{1/2} \int_0^\infty F(t) \cos xt\, dt, \text{ and} \qquad (1)$$

$$F(t) = \left(\frac{2}{\pi}\right)^{1/2} \int_0^\infty f(x) \cos xt\, dx,$$

then these functions are related to one another by the following formula:

$$(\beta)^{1/2}\left[\tfrac{1}{2} f(0) + \sum_{n=1}^\infty f(n\beta)\right] = (\alpha)^{1/2}\left[\tfrac{1}{2} F(0) + \sum_{n=1}^\infty F(n\alpha)\right], \qquad (2)$$

where α and β are positive numbers such that $\alpha\beta = 2\pi$.

---

*Mém de l'Acad. des Sci., Vol. 6, 1827, p. 592. See also: Journ. Ecole Poly., Cah., 19, 1823.

†See comment by H. Burkhardt: *Trigonometrische Reihen und Integrale*. Article II A 12 in *Encyklopädie der Math. Wissenschaften*, pp. 1339-1342.

§The following references may be cited: E. C. Titchmarsh: *Introduction to the Theory of Fourier Series*. Oxford, 1937, pp. 60-62. S. Bochner: *Vorlesungen über Fourische Integrale*. Leipzig, 1932, pp. 33-38; 203-208. H. Burkhardt: *loc. cit.* A. L. Cauchy: *Oeuvres*, 1st ser., Vol. 3, pp. 308 *et seq.* G. Landsberg: *Journal für Math.*, Vol. 111, 1893, pp. 234-253. E. H. Linfoot: *Journal London Math. Soc.*, Vol. 4 1928, pp. 54-61; L. J. Mordell: *Ibid*, Vol. 4, 1928, pp. 285-291; S. Borgen: *Ibid*, Vol. 19, 1944, pp. 213-219; R. P. Boas, Jr., *Ibid.*, Vol. 21, 1946, pp. 102-105. R. Goldberg and R. S. Varga: *Duke Math. Journal*, Vol. 24, 1956, pp. 553-560.

This formula holds if $f(x)$ and $F(x)$ are continuous functions, which satisfy the additional criteria that assure the existence of the integrals in (1). This restrictive condition on the functions has been modified in various ways, such as letting $F(x)$ be of limited variation in $(0, \infty)$, and the formula has been appropriately generalized. The reader is referred to the literature for a detailed discussion of these extensions.

To establish Poisson's formula under general conditions is a matter of considerable difficulty. But it can be proved for the restrictive assumption that both functions are continuous by the following argument, which is adapted from a proof given by E. H. Linfoot.

Let $g(t)$ be a function defined by the following sum:

$$g(t) = \sum_{n=-\infty}^{\infty} G(2n\pi + t), \tag{3}$$

which is assumed to be uniformly convergent in the interval $|t| \leqslant \pi$.

Then $g(t)$ can be expanded in the following Fourier series
$$g(t) = \tfrac{1}{2} A_0 + A_1 \cos t + A_2 \cos 2t + A_3 \cos 3t + \cdots , \tag{4}$$
$$+ B_1 \sin t + B_2 \sin 2t + B_3 \sin 3t + \cdots ,$$

where we have

$$A_m = \frac{1}{\pi} \int_{-\pi}^{\pi} g(s) \cos ms\, ds, \quad B_m = \frac{1}{\pi} \int_{-\pi}^{\pi} g(s) \sin ms\, ds. \tag{5}$$

Hence we can write,

$$g(t) = \frac{1}{2\pi} \int_{-\pi}^{\pi} g(s)\, ds + \frac{1}{\pi} \sum_{m=1}^{\infty} \int_{-\pi}^{\pi} \cos m(t-s)\, g(s)\, ds,$$

$$= \frac{1}{2\pi} \sum_{n=-\infty}^{\infty} \int_{-\pi}^{\pi} G(2n\alpha + s)\, ds$$

$$+ \frac{1}{\pi} \sum_{m=1}^{\infty} \sum_{n=-\infty}^{\infty} \int_{-\pi}^{\pi} \cos m(t-s)\, G(2n\pi + s)\, ds,$$

which is possible through the assumption of the uniform convergence of series (3).

Let us now make the transformation:

$$2n\pi + s = w, \quad t - s = t - w - 2n\pi.$$

We thus get

$$g(t) = \frac{1}{2\pi} \sum_{n=-\infty}^{\infty} \int_{(2n-1)\pi}^{(2n+1)\pi} G(w) \, dw$$

$$+ \frac{1}{\pi} \sum_{m=1}^{\infty} \sum_{n=-\infty}^{\infty} \int_{(2n-1)\pi}^{(2n+1)\pi} \cos m(t - w - 2n\pi) \, G(w) \, dw,$$

$$= \frac{1}{2\pi} \int_{-\infty}^{\infty} G(w) \, dw + \frac{1}{\pi} \sum_{m=1}^{\infty} \int_{-\infty}^{\infty} \cos m(t - w) \, G(w) \, dw, \tag{6}$$

$$= \frac{1}{2\pi} \sum_{m=-\infty}^{\infty} \int_{-\infty}^{\infty} \cos m(t - w) \, G(w) \, dw.$$

If we now let $t = 0$, we get

$$\sum_{n=-\infty}^{\infty} G(2n\pi) = \frac{1}{2\pi} \sum_{m=-\infty}^{\infty} \int_{-\infty}^{\infty} \cos mw \, G(w) \, dw. \tag{7}$$

Replacing $G(x)$ by $f(x\beta/2\pi)$ and changing the summation variable from m to n, we have

$$\sum_{n=-\infty}^{\infty} f(n\beta) = \frac{1}{2\pi} \sum_{m=-\infty}^{\infty} \int_{-\infty}^{\infty} \cos nw \, f\left(\frac{\beta w}{2\pi}\right) dw. \tag{8}$$

By means of the transformation: $w = \alpha s$, where $\alpha\beta = 2\pi$, we get

$$\sum_{n=-\infty}^{\infty} f(n\beta) = \frac{\alpha}{2\pi} \sum_{n=-\infty}^{\infty} \int_{-\infty}^{\infty} \cos \alpha \, ns \, f(s) \, ds. \tag{9}$$

If we now define $F(n)$ as the cosine transform $f(s)$, that is,

$$F(n) = \frac{1}{(2\pi)^{1/2}} \int_{-\infty}^{\infty} \cos ns \, f(s) \, ds, \tag{10}$$

we can write (9) as follows:

$$\sum_{n=-\infty}^{\infty} f(n\beta) = \frac{\alpha}{(2\pi)^{1/2}} \sum_{n=-\infty}^{\infty} F(n\alpha), \tag{11}$$

which, since $\alpha\beta = 2\pi$, reduces to

$$(\beta)^{1/2} \sum_{n=-\infty}^{\infty} f(n\beta) = (\alpha)^{1/2} \sum_{n=-\infty}^{\infty} F(n\alpha). \tag{12}$$

This somewhat more general statement of Poisson's formula is observed to include (2) if we assume that $f(x)$ is an even function, that is, $f(-x) = f(x)$.

The following examples will illustrate the application of Poisson's formula to the summation of series:

Example 1. One of the celebrated problems is furnished by the case where $f(x) = e^{-x^2}$. This problem appears to have originated with Poisson and attracted the attention of K. G. J. Jacobi and other analysts. Jacobi found a deep-seated relationship between the sums involved and the expansions of Theta functions.*

Using equations (1) and (2), we compute

$$F(t) = \left(\frac{2}{\pi}\right)^{1/2} \int_0^{\infty} e^{-x^2} \cos xt \, dt = \tfrac{1}{2} (2)^{1/2} e^{-t^2/4}.$$

Substituting appropriate values in (2) we then obtain the following sum:

$$\sum_{n=0}^{\infty} e^{-n^2\beta^2} = \frac{1}{2} (\pi)^{1/2} + \tfrac{1}{2} + R(\beta), \tag{13}$$

where we abbreviate:

$$R(\beta) = \frac{1}{\beta} (\pi)^{1/2} \sum_{n=1}^{\infty} e^{-n^2\pi^2/\beta^2}.$$

---

*See, for example, E. T. Whittaker and G. N. Watson: *Modern Analysis*, 4th ed., Cambridge, 1927, pp. 474-476; also, p. 124.

We observe that $R(\beta)$ rapidly diminishes as β approaches zero. When $\rho = 1$, $R(\beta) = 0.00009\ 16769$ and when $\beta = \frac{1}{2}$, $R(\beta)$ is less than 10^{-10}.

Example 2. We shall let $f(x) = \sin x/x$. In this case we have

$$F(n\alpha) = F(2n\pi/\beta) = \tag{14}$$

$$\left(\frac{2}{\pi}\right)^{1/2} \int_0^\infty \frac{\sin x}{x} \cos \frac{2\pi n x}{\beta}\, dx = \begin{cases} (\frac{1}{2}\pi)^{1/2}, & \beta > 2n\pi, \\ 0, & \beta < 2n\pi, \\ (\frac{1}{8}\pi)^{1/2}, & \beta = 2n\pi. \end{cases}$$

Assuming that β lies between 0 and π, we have from (2)

$$\sum_{n=0}^\infty \sin \frac{n\beta}{n} = (2\pi)^{1/2}\left[\frac{1}{2} F(0)\right] + \frac{1}{2} = \frac{1}{2}(\pi + \beta), 0 < \beta < \pi. \tag{15}$$

Example 3. Prove that,

$$\sum_{n=1}^\infty \frac{n \sin n\beta}{a^2 + n^2} = \frac{\pi}{2} \frac{\sinh a(\pi - \beta)}{\sinh a\pi}, 0 < \beta < 2\pi.$$

Solution: Let us now write,

$$f(x) = \frac{x \sin x}{b^2 + x^2}.$$

We then have

$$F(t) = \left(\frac{2}{\pi}\right)^{1/2} \int_0^\infty f(x) \cos xt\, dx,$$

from which it is readily shown that we have the following values:

$$F(t) = \begin{cases} \left(\frac{\pi}{2}\right)^{1/2} e^{-b} \cosh bt, & t < 1, \\ -\left(\frac{\pi}{2}\right)^{1/2} e^{-bt} \sinh b, & t > 1, \\ \frac{1}{2}\left(\frac{\pi}{2}\right)^{1/2} e^{-2b}, & t = 1. \end{cases}$$

We thus see, when $t = 2n\pi/\beta$, and β lies between 0 and 2π, that

$$F(0) = \left(\frac{\pi}{2}\right)^{1/2} e^{-b}, \quad F(2n\pi/\beta) = -\left(\frac{\pi}{2}\right)^{1/2} e^{-2n\pi b/\beta} \sinh b,$$

from which we compute

$$\sum_{n=1}^{\infty} F(2n\pi/\beta) = - \left(\frac{\pi}{2}\right)^{1/2} \frac{1}{2} \frac{e^{-\pi b/\beta}}{\sinh (\pi b/\beta)} \sinh b.$$

If we now make appropriate substitutions in equation (2), we have

$$(\beta)^{1/2} \sum_{x=1}^{\infty} \frac{x \sin x\beta}{b^2 + x^2\beta^2} = \frac{1}{(\beta)^{1/2}} \sum_{x=0}^{\infty} \frac{x \sin x\beta}{a^2 + x^2} =$$

$$\frac{\pi}{(\beta)^{1/2}} \left[\tfrac{1}{2} e^{-a} - \tfrac{1}{2} \sinh a\beta \frac{e^{-ax}}{\sinh a\pi} \right],$$

where we have written $b = a\beta$.

The expression in brackets is readily shown to reduce to

$$\frac{\sinh a(\pi - \beta)}{2 \sinh a\pi},$$

from which the required sum follows as an immediate consequence.

8. *Comparison of the Euler-Maclaurin and Poisson Formulas*. It will be of interest to observe the resemblance and the difference between the Poisson formula and the Euler-Maclaurin formula in which the summation has been extended to infinity. Referring to equation (1) of Section 8, Chapter 2, we let $a = 0$ and $n \to \infty$. If the sum is to converge it is necessary that $f(a + nd) \to 0$ as $n \to \infty$.

Observing that we can write

$$f^{(p)}(a + nd) - f^{(p)}(a) = \int_0^{\infty} D^{p+1} f(x) \, dx, \qquad (1)$$

where D is the derivative symbol, we obtain the following expression for the Euler-Maclaurin formula:

$$\tfrac{1}{2} f(0) + \sum_{x=1}^{\infty} f(xd) = \frac{1}{d} \int_0^{\infty} f(x) \, dx + R, \qquad (2)$$

where R is the sum

$$R = \sum_{n=1}^{\infty} \int_0^{\infty} (-1)^{n-1} \frac{B_n d^{2n-1}}{(2n)!} D^{2n} f(x) \, dx. \qquad (3)$$

In Poisson's formula, equation (2) of the preceding section, we observe that the right hand member can be written:

$$\left(\frac{2\alpha}{\pi}\right)^{1/2}\left[\frac{1}{2}\int_0^\infty f(x)\ dx + \sum_{n=1}^\infty \int_0^\infty f(x)\cos nx\alpha\ dx\right].\qquad (4)$$

If we replace β by d, α by $2\pi/d$, and change the variable of integration in the second member from x to $x\alpha$, then (4) can be written:

$$\frac{1}{d}\int_0^\infty f(x)\ dx + \frac{1}{\pi}\sum_{n=1}^\infty \int_0^\infty f\left(\frac{xd}{2\pi}\right)\cos nx\ dx.\qquad (5)$$

We can thus write Poisson's formula, comparable to (2), as follows:

$$\tfrac{1}{2}\,f(0) + \sum_{x=1}^\infty f(xd) = \frac{1}{d}\int_0^\infty f(x)\ dx + R',\qquad (6)$$

where R' is the sum

$$R' = \frac{1}{\pi}\sum_{n=1}^\infty \int_0^\infty f\left(\frac{xd}{2\pi}\right)\cos nx\ dx.\qquad (7)$$

A striking difference between the two formulas is shown by the following example:

Example 1. Evaluate the sums:

$$(a)\ \sum_{x=0}^\infty e^{-xd};\qquad (b)\ \sum_{x=0}^\infty e^{-x^2d^2}.$$

Solution: (a) Let us first observe that if $G(D)$ is a linear operator which does not contain x explicitly, then

$$G(D) \to e^{ax} = G(a)\ e^{ax},$$

where $\to$ is the symbol of operation. Referring to equation (3), we observe that

$$G(D) = \sum_{n=1}^\infty (-1)^{2n-1}\frac{B_n d^{2n-1}}{(2n)\,!}\,D^{2n} = \frac{1}{d}\left[\frac{dD}{e^{dD}-1} - 1 + \tfrac{1}{2}dD\right].$$

Hence, by (3), if $f(x) = e^{-x}$, we have

$$\sum_{x=0}^{\infty} e^{-xd} = \tfrac{1}{2} + \frac{1}{d} - \frac{1}{d} \int_0^{\infty} \left[\frac{d}{e^{-d} - 1} - 1 - \tfrac{1}{2}d \right] e^{-x} \, dx, \quad (8)$$

$$= 1 + \frac{1}{e^d - 1}.$$

In this example, we have interchanged the order of summation and integration, since the series is uniformly convergent.

If we solve this problem by Poisson's formula, equation (6), we obtain

$$\sum_{x=0}^{\infty} e^{-xd} = \tfrac{1}{2} + \frac{1}{d} + \frac{1}{\pi} \sum_{n=1}^{\infty} \frac{\alpha}{\alpha^2 + n^2}, \text{ where } \alpha = d/2\pi.$$

In other words, we have the value of one series in terms of another. If one assumes the answer just obtained in (8), then one finds the following:

$$\sum_{n=1}^{\infty} \frac{\alpha}{\alpha^2 + n^2} = \frac{\pi\alpha \coth \pi\alpha - 1}{2\alpha}. \quad (9)$$

(b) We have already solved this problem by Poisson's formula in Example 1 of the preceding section, where it was shown that R', in the notation of this section, had the following form:

$$R' = \frac{1}{d} (\alpha)^{1/2} \sum_{n=1}^{\infty} e^{-n^2 a^2 / d^2}.$$

But if we make use of the Euler-Maclaurin formula (2), we encouter a curious difficulty. Each term in R, equation (3), is zero. To see this, let us write

$$D^p e^{-x^2} = (-1)^p H_p(x) e^{-x^2},$$

where $H_p(x)$ is the Hermite polynomial of degree p. If p is odd, then $H_p(p)$ is an odd function and vanishes when $x = 0$. Since e^{-x^2} dominates any polynomial, all derivatives also vanish at infinity.

The mystery is resolved, however, if we interchange the order of summation and integration in (3) and thus consider the series:

$$S = \sum_{n=1}^{\infty} (-1)^{n-1} \frac{B_n d^{2n-1}}{(2n)!} D^{2n} e^{-x^2}.$$

In order to investigate the convergence of this series, let us apply the radical test [See (4), Section 2]. We first observe that

$$B_n \sim 2(2n)\,!/(2\pi)^{2n}.$$

Since e^{-x^2} is an entire function of genus 2, the nth root of its nth derivative is of order $n^{\frac{1}{2}}$ (see, Theorem 1, Section 6). Thus the nth root of the nth term in series S is of order $kn^{\frac{1}{2}}$, where k is a constant, and by the radical test the series is divergent. Thus the integral of the sum is not equal to the sum of the integrals as was the case in problem (a) above.

Example 2. Find the value of the following series for θ between 0 and π:

$$S(\theta) = \sin \theta - \frac{\sin 3\theta}{3^2} + \frac{\sin 5\theta}{5^2} - \frac{\sin 7\theta}{7^2} + \cdots \cdot \qquad (10)$$

Solution: In this case we choose

$$f(x) = \frac{\sin x\theta \sin \frac{1}{2}\pi x}{x^2},$$

and observing the value of the following integral:

$$\int_0^\infty \frac{\sin px \sin qx}{x^2}\,dx = C_q, \qquad (11)$$

where $C = \frac{1}{2}\pi$ if $q < p$, $C = 0$ if $q > p$, $C = \frac{1}{4}\pi$ if $q = p$.

We thus compute

$$\int_0^\infty f(x)\,dx = \begin{cases} \frac{1}{2}\pi\theta, & 0 \leqslant \theta \leqslant \frac{1}{2}\pi, \\ \frac{1}{4}\pi^2, & \frac{1}{2}\pi \leqslant \theta \leqslant \pi. \end{cases} \qquad (12)$$

Observing also that $f(\theta) = \frac{1}{2}\pi\theta$, we then have

$$\frac{1}{4}\pi\theta + S(\theta) = \begin{cases} \frac{1}{2}\pi\theta \\ \frac{1}{4}\pi^2 \end{cases} + R \text{ or } R'. \qquad (13)$$

In the case of R it is readily shown by expanding $f(x)$ in a power series that all odd derivatives are zero at $x = 0$. With more difficulty it can be shown that the derivatives vanish also at infinity. Hence $R = 0$.

In the case of R, we must evaluate the integral

$$\int_0^\infty f\left(\frac{x}{2\pi}\right) \cos nx\,dx.$$

To do this cos nx is combined with one of the sine factors in $f(x)$ to give the sum of two sine terms and formula (11) is then used to evaluate the integral. It will be found that $R' = 0$.

Thus, from (13), we obtain finally the desired sum as follows:

$$S(\theta) = \begin{cases} \tfrac{1}{2}\pi\theta - \tfrac{1}{4}\pi\theta = \tfrac{1}{4}\pi\theta & 0 \leqslant \theta \leqslant \tfrac{1}{2}\pi, \\ \tfrac{1}{4}\pi^2 - \tfrac{1}{4}\pi\theta = \tfrac{1}{4}\pi(\pi - \theta), & \tfrac{1}{2}\pi \leqslant \theta \leqslant \pi. \end{cases} \tag{14}$$

PROBLEMS

1. Show that
$$\sin\theta + \frac{\sin 3\theta}{3} + \frac{\sin 5\theta}{5} + \cdots = \tfrac{1}{4}\pi, \; 0 < \theta < \pi.$$

2. Prove that the value of the following series is $\tfrac{1}{2}\theta$ for $0 < \theta < \pi$:
$$\sin\theta - \frac{\sin 2\theta}{2} + \frac{\sin 3\theta}{3} - \frac{\sin 4\theta}{4} + \cdots .$$

3. Show that
$$\sum_{x=1}^{\infty} \frac{\sin^2 x\theta}{x^2} = \tfrac{1}{2}\theta(\pi - \theta), \; 0 < 0 \leqslant \pi.$$

4. Given the following integral
$$\int_0^{\infty} \frac{\cos x}{x^2 + b^2}\, dx = \frac{\pi}{2b} e^{-b},$$
find the value of the sum:
$$S(a) = \sum_{x=1}^{\infty} \frac{1}{x^2 + a^2}.$$

5. Evaluate the integral
$$\int_0^{\infty} \frac{1}{x^2} [\sin px \sin qx \cos rx]\, dx.$$
Letting $p = \theta/2\pi$, $q = \tfrac{1}{4}$, $r = n$, show that R' in Example 2 is zero.

6. Show that
$$\sum_{n=0}^{\infty} \frac{\sin n\pi x \sin n\pi t}{n^2} = \tfrac{1}{2}\pi^2 \begin{cases} x(1 - t), & 0 \leqslant x \leqslant t, \\ t(1 - x), & t \leqslant x \leqslant 1. \end{cases}$$

TABLE OF FINITE SUMS

A. General Forms.

1. $\sum \Delta f(x) = f(x)$.

2. $\sum [\Delta f(x) \pm \Delta g(x) \pm \cdots] = \sum \Delta f(x) \pm \sum \Delta g(x) \pm \cdots$
$$= f(x) \pm g(x) \pm \cdots .$$

3. $\sum u_x \, \Delta v_x = u_x \, v_x - \sum v_{x+1} \, \Delta u_x$.

4. $\sum u_x \, v_x = u_x \sum v_x - \sum \left(\Delta u_x \sum v_{x+1} \right)$.

5. $\sum \dfrac{u_{x+1} \, \Delta v_x + v_x \, \Delta u_x}{u_x \, u_{x+1} \, v_x \, v_{x+1}} = - \dfrac{1}{u_x \, v_x}$.

6. $\sum (u_{x+1} - u_{x-n}) u_x^{(n)} = u_x^{(n+1)}$,

where $u_x^{(n)} = u_x \, u_{x-1} \, u_{x-2} \cdots u_{x-n+1}$.

7. $\sum \dfrac{v_x \, \Delta u_x - u_x \, \Delta v_x}{v_x \, v_{x+1}} = \dfrac{u_x}{v_x}$.

B. Forms Involving Rational Coefficients and Powers of x.

1. $\sum a = ax$.

2. $\sum x^{(n)} = \dfrac{x^{(n+1)}}{n + 1}, \quad n \neq -1$.

3. $\sum (a + bx)^{(n)} = \dfrac{(a + bx)^{(n+1)}}{b \, (n + 1)}$, where $(a + bx)^{(n)}$
$$= (a + bx) \, (a + bx - b) \cdots (a + bx - nb + b).$$

4. $\sum \dfrac{1}{x} = \Psi(x)$.

5. $\sum \dfrac{1}{x^2} = -\Psi'(x).$

6. $\sum \dfrac{1}{x^3} = \dfrac{1}{2!}\,\Psi''\,(x).$

7. $\sum \dfrac{1}{x^n} = \dfrac{(-1)^{n+1}}{(n-1)!}\,\Psi^{(n-1)}(x).$

8. $\sum (cx+d)^{(-n)} = -(cx+d)^{(-n+1)},$ where $(cx+d)^{(-n)} =$
$$1/[(cx+d+c)(cx+d+2c)\cdots(cx+d+nc)].$$

9. $\sum x(x+a)(x+2a) = \dfrac{x^{(2)}(x+2a)^{(2)}}{4}.$

10. $\sum x \cdot x! = x!.$

11. $\sum \dfrac{x-1}{x!} = -\dfrac{1}{(x-1)!}.$

12. $\sum \dfrac{x+a-1}{(x+a)!} = -\dfrac{1}{(x+a-1)!},$ a an integer; otherwise:
$$\sum \dfrac{(x+a-1)}{\Gamma(x+a+1)} = -\dfrac{1}{\Gamma(x+a)}.$$

13. $\sum (2x+1) = x^2.$

14. $\sum (3x^2+3x+1) = x^3.$

15. $\sum (4x^3+6x^2+4x+1) = x^4.$

16. $\sum x = \dfrac{x(x-1)}{2}.$

17. $\sum \begin{bmatrix} x \\ n \end{bmatrix} = \begin{bmatrix} x \\ n+1 \end{bmatrix}$, where $\begin{bmatrix} x \\ n \end{bmatrix} = \dfrac{x!}{n!(x-n)!}$.

18. $\sum x^{(-n)} = \dfrac{x^{(-n+1)}}{-n+1}$, $n \neq 1$.

19. $\sum \begin{pmatrix} x \\ 1 \end{pmatrix} \begin{pmatrix} x \\ 3 \end{pmatrix} = \begin{pmatrix} x \\ 1 \end{pmatrix} \begin{pmatrix} x \\ 4 \end{pmatrix} - \begin{pmatrix} x+1 \\ 5 \end{pmatrix}$, where $\begin{pmatrix} x \\ n \end{pmatrix} = \dfrac{x^{(n)}}{n!}$.

20. $\sum \dfrac{(x-1)^{(x-1)}}{(x+1)^{(x-1)}} = -(x+1)\dfrac{(x-1)^{(x-1)}}{(x+1)^{(x-1)}}$.

21. $\sum \left[\dfrac{(x-1)^{(x-1)}}{(x+1)^{(x-1)}}\right]^2 (2x+1) = -(x+1)^2 \left[\dfrac{(x-1)^{(x-1)}}{(x+1)^{(x-1)}}\right]^2$.

22. $\sum \dfrac{1}{x(x+1)} = -\dfrac{1}{x}$.

23. $\sum \dfrac{2x+1}{x^2(x+1)^2} = -\dfrac{1}{x^2}$.

24. $\sum \dfrac{3x^2+3x+1}{x^3(x+1)^3} = -\dfrac{1}{x^3}$.

25. $\sum \dfrac{{}_nC_1 x^{n-1} + {}_nC_2 x^{n-2} + \cdots + 1}{x^n(x+1)^n} = -\dfrac{1}{x^n}$,

$\qquad\qquad\qquad$ where ${}_nC_r = \dfrac{n!}{r!\,(n-r)!}$.

26. $\sum \dfrac{1}{x\,(x+1)(x-1)} = -\dfrac{1}{2x^{(2)}}$.

27. $\sum \dfrac{1}{x\,(x+1)(x-1)(x-2)} = -\dfrac{1}{3x^{(3)}}$.

28. $\sum \dfrac{1}{x\,(x+1)(x-1)\cdots(x-n+1)} = -\dfrac{1}{nx^{(n)}}$.

29. $\sum (x+1)^{(m+1)} (x+1)^{(n+1)} \times$

$$\left[\frac{(x+1)^2 - (x-m+1)(x-n+1)}{(x-m+1)(x-n+1)(x+1)^2} \right] = x^{(m)} x^{(n)}.$$

C. Forms Involving a^x and 2^x.

1. $\sum a^x = \dfrac{a^x}{a-1}.$

2. $\sum 2^x = 2^x.$

3. $\sum x 2^x = 2^x (x-2).$

4. $\sum x a^x = \dfrac{a^x}{a-1} \left[x - \dfrac{a}{a-1} \right], a \neq 1.$

5. $\sum 2^{ax} = \dfrac{2^{ax}}{2^a - 1}.$

6. $\sum x 2^{ax} = \dfrac{2^{ax}}{2^a - 1} \left[x - \dfrac{2^a}{2^a - 1} \right].$

7. $\sum \dfrac{2\,a^{f(x)}}{1 - a^{f(x+1)}} = - \dfrac{1 + a^{f(x)}}{1 - a^{f(x)}},$ where $f(x) = 2^{x-1}.$

8. $\sum x^{(n)} a^x = x^{(n)} \dfrac{a^x}{a-1} - \dfrac{na}{a-1} \sum a^x x^{(n-1)}, a \neq 1.$

9. $\sum \dfrac{a^x}{(1-a^x)(1-a^{x+1})} = - \dfrac{1}{(1-a)(1-a^x)}, a \neq 1.$

10. $\sum 2^{-x} = - 2^{(-x+1)}.$

11. $\sum (-1)^x a^x = (-1)^{x+1} \dfrac{a^x}{a+1}.$

12. $\displaystyle\sum \frac{x}{2^x} = -\frac{x+1}{2^{x-1}}.$

13. $\displaystyle\sum (x+2)2^x = x\,2^x.$

14. $\displaystyle\sum (x^2 + 4x + 2)2^x = x^2 2^x.$

15. $\displaystyle\sum (x^3 + 2\cdot 3x^2 + 2\cdot 3x + 2)2^x = x^3 2^x.$

16. $\displaystyle\sum (x^n + 2nx^{n-1} + \cdots + 2)2^x = x^n 2^x.$

17. $\displaystyle\sum \frac{2^x(x-3)}{x(x+1)(x-1)} = \frac{2^x}{x^{(2)}};\quad \sum \frac{2^x(x-3)}{x^{(2)}(x+1)} = \frac{2^x}{x^{(2)}}.$

18. $\displaystyle\sum \frac{2^x(x-1)}{x(x+1)} = \frac{2^x}{x}.$

19. $\displaystyle\sum \frac{2^x(x - 2\cdot 3 + 1)}{x(x+1)(x-1)(x-2)} = \frac{2^x}{x^{(3)}};\quad \sum \frac{2^x(x-5)}{x^{(3)}(x+1)} = \frac{2^x}{x^{(3)}}.$

20. $\displaystyle\sum \frac{2^x(x - 2\cdot 4 + 1)}{x(x+1)(x-1)(x-2)(x-3)} = \sum \frac{2^x(x-7)}{x^{(4)}(x+1)} = \frac{2^x}{x^{(4)}}.$

21. $\displaystyle\sum \frac{2^x(x - 2n + 1)}{x^{(n)}(x+1)} = \frac{2^x}{x^{(n)}}.$

22. $\displaystyle\sum \frac{2^x(x^2 - 2x - 1)}{x^2(x+1)^2} = \frac{2^x}{x^2}.$

23. $\displaystyle\sum \frac{2^x(x^3 - 3x^2 - 3x - 1)}{x^3(x+1)^3} = \frac{2^x}{x^3}.$

24. $\displaystyle\sum \frac{2^x(x^n - nx^{n-1} - \cdots - 1)}{x^n(x+1)^n} = \frac{2^x}{x^n}.$

25. $\sum \dfrac{2^x(A + x - 1)}{(A + x + 1)(A + x)} = \dfrac{2^x}{A + x}.$

26. $\sum \dfrac{2^x(A + x^2 - 3x)}{(A + x^2 + x)(A + x^2 - x)} = \dfrac{2^x}{A + x^{(2)}}.$

27. $\sum \dfrac{2^x \cdot x}{(2 + x)(1 + x)} = \dfrac{2^x}{x + 1}.$

28. $\sum \dfrac{2^x(x^2 - 3x + 1)}{(x^2 + x + 1)(x^2 - x + 1)} = \dfrac{2^x}{1 + x^{(2)}}.$

29. $\sum \dfrac{(1 + x)2^x}{(x + 2)(x + 3)} = \dfrac{2^x}{x + 2}.$

30. $\sum 2^x(x - 1)(x + 2) = 2^x(x - 1)^{(2)}.$

D. Forms Involving e^x.

1. $\sum e^x[(x + 1)e - x] = xe^x.$

2. $\sum xe^x[(x + 1)e - (x - 1)] = x^{(2)}e^x.$

3. $\sum e^x[(x + 1)(e - 1) + e] = xe^x + e^x = e^x(x + 1).$

4. $\sum \dfrac{(e - 2) + x(e - 1)}{(x + 2)(x + 1)} e^x = \dfrac{e^x}{x + 1}.$

5. $\sum e^x = \dfrac{e^x}{e - 1}.$

6. $\sum e^{2x} = \dfrac{e^{2x}}{e^2 - 1}.$

7. $\sum e^{nx} = \dfrac{e^{nx}}{e^n - 1}.$

E. Forms Involving Trigonometric Functions.

1. $\displaystyle\sum \sin(ax + b) = \frac{-\cos(ax + b - a/2)}{2 \sin a/2}$

$\qquad\qquad\qquad = \dfrac{\sin(ax + b - a/2 - \pi/2)}{2 \sin a/2}.$

2. $\displaystyle\sum \cos(ax + b) = \frac{\sin(ax + b - a/2)}{2 \sin a/2}$

$\qquad\qquad\qquad = \dfrac{\cos(ax + b - a/2 - \pi/2)}{2 \sin a/2}.$

3. $\displaystyle\sum x \sin x = \frac{-x \cos(x - 1/2)}{2 \sin 1/2} + \frac{\sin x}{(2 \sin 1/2)^2}.$

4. $\displaystyle\sum x \cos x = \frac{x \sin(x - 1/2)}{2 \sin 1/2} + \frac{\cos x}{(2 \sin 1/2)^2}.$

5. $\displaystyle\sum \frac{1}{\cos ax \cos (x + 1)a} = \csc a \tan ax.$

6. $\displaystyle\sum \frac{1}{\sin ax \sin (x + 1)a} = -\sec a \cot ax.$

7. $\displaystyle\sum \frac{\sin (ax + a/2)}{\cos ax \cos (x + 1)a} = \frac{\sec ax}{2 \sin a/2}.$

8. $\displaystyle\sum \frac{\cos (ax + a/2)}{\sin ax \sin (x + 1)a} = \frac{-\csc ax}{2 \sin a/2}.$

9. $\displaystyle\sum \frac{\tan (\theta/2^{x+1})}{\cos (\theta/2^x)} = - \tan (\theta/2^x).$

10. $\displaystyle\sum a^x \cos \theta x = a^x \left\{ \frac{a \cos (x - 1) \theta - \cos \theta x}{a^2 - 2a \cos \theta + 1} \right\}.$

11. $\displaystyle\sum a^x \sin \theta x = a^x \left\{ \frac{a \sin (x - 1) \theta - \sin \theta x}{a^2 - 2a \cos \theta + 1} \right\}.$

12. $\displaystyle\sum 2^{x+2} \sin \frac{\theta}{2^{x+1}} \left(\sin \frac{\theta}{2^{x+2}} \right)^2 = 2^x \sin \frac{\theta}{2^x}.$

13. $\displaystyle\sum 2^{2x+2}\left(\sin\frac{\theta}{2^{x+1}}\right)^4 = 2^{2x}\left(\sin\frac{\theta}{2^x}\right)^2.$

14. $\displaystyle\sum 2^x \tan\frac{\theta}{2^x}\left(\tan\frac{\theta}{2^{x+1}}\right)^2 = -2^x \tan\frac{\theta}{2^x}.$

15. $\displaystyle\sum (-1)^{x+1}2^{x+3}\sin\frac{\theta}{2^{x+2}}\left(\cos\frac{\theta}{2^{x+2}}\right)^3 = (-2)^x \sin\frac{\theta}{2^x}.$

16. $\displaystyle\sum\frac{(\sin\theta/2^{x+2})^2}{2^{x-1}\sin\theta/2^x} = -\frac{1}{2^x\sin\theta/2^x}.$

17. $\displaystyle\sum\frac{\tan\theta/2^{x+1}}{2^{x+1}} = \frac{1}{2^x\tan\theta/2^x}.$

18. $\displaystyle\sum\left(\frac{1}{2^{2x+1}} - \frac{\tan^2\theta/2^{x+1}}{2^{2x+2}}\right) = \left(\frac{1}{2^x\tan\theta/2^x}\right)^2.$

19. $\displaystyle\sum\frac{\sin\theta/2^{x+1}}{(2^{x+1}\cos\theta/2^{x+1})^3} = \frac{\cos\theta/2^x}{(2^x\sin\theta/2^x)^3}.$

20. $\displaystyle\sum\frac{1}{\sin 2^{x+1}\theta} = -\cot 2^x\theta.$

21. $\displaystyle\sum\frac{1}{(2^{x+1}\cos\theta/2^{x+1})^2} = -\frac{1}{(2^x\sin\theta/2^x)^2}.$

22. $\displaystyle\sum\frac{2\cos\theta(-1)^{x+1}}{\sin(a+\theta x)\sin[a+\theta(x+2)]}$

$$= \frac{(-1)^x}{\sin(a+\theta x)\sin[a+(x+1)\theta]}.$$

23. $\displaystyle\sum\frac{2\cos\theta(-1)^{x+1}}{\cos(a+\theta x)\cos[a+\theta(x+2)]}$

$$= \frac{(-1)^x}{\cos(a+\theta x)\cos[a+\theta(x+1)]}.$$

24. $\displaystyle\sum\frac{\csc ax\,\csc(ax+a)}{\sec\frac{1}{2}(2ax+a)} = -\frac{\csc ax\,\csc\frac{1}{2}a}{2}.$

25. $\displaystyle\sum \frac{\sec (ax + a) \sec ax}{\csc \frac{1}{2}(2 ax + a)} = \frac{\sec ax \csc \frac{1}{2}a}{2}.$

26. $\displaystyle\sum \tan (ax + a) \tan ax = \frac{\tan ax}{\tan a} - x.$

27. $\displaystyle\sum \cot (ax + a) \cot ax = - \cot a \cot ax - x.$

28. $\displaystyle\sum \sin (2ax - a) = \begin{cases} - \dfrac{\cos^2 (ax - a)}{\sin a}, \\[2ex] \dfrac{\sin ax \sin (ax - 2a)}{\sin a}. \end{cases}$

29. $\displaystyle\sum \sin (2ax + a) = \frac{\sin^2 ax}{\sin a}.$

30. $\displaystyle\sum x \sin (ax + b) = \frac{-x \cos (ax + b - a/2)}{2 \sin a/2} + \frac{\sin (ax + b)}{(2 \sin a/2)^2}.$

31. $\displaystyle\sum x \cos (ax + b) = \frac{x \sin (ax + b - a/2)}{2 \sin a/2} + \frac{\cos (ax + b)}{(2 \sin a/2)^2}.$

32. $\displaystyle\sum x^{(2)} \sin (ax + b) = - \frac{x^{(2)} \cos (ax + b - a/2)}{2 \sin a/2}$

$$+ \frac{1}{\sin a/2}\left[\frac{x \sin (ax + b)}{2 \sin a/2} + \frac{\cos (ax + b + a/2)}{(2 \sin a/2)^2} \right].$$

33. $\displaystyle\sum x^{(2)} \cos (ax + b) = \frac{x^{(2)} \sin (ax + b - a/2)}{2 \sin a/2}$

$$+ \frac{1}{\sin a/2}\left[\frac{x \cos (ax + b)}{2 \sin a/2} - \frac{{'}\sin (ax + b + a/2)}{(2 \sin a/2)^2} \right].$$

34. $\displaystyle\sum \cos (2ax + a) = \frac{\sin 2ax}{2 \sin a}$

F.   Forms Involving Logarithms.

1. $\sum \log x = \log \Gamma(x)$.

2. $\sum [x \log x + \log \Gamma(x)] = (x - 1) [\log \Gamma(x)]$.

2a. $\sum \left[x^p \log \Gamma(x) + \dfrac{1}{p + 1} B_{p+1}(x + 1) \log x \right]$

$$= \frac{1}{p + 1} [B_{p+1}(x) \log \Gamma(x)],$$

where $B_p(x)$ is the pth Bernoulli polynomial.

3. $\sum \dfrac{\log (\cot 2^x \theta)}{2^{x+1}} = \dfrac{\log (2 \sin 2^x \theta)}{2^x}$.

4. $\sum \dfrac{\log (\tan 2^x \theta)}{2^{x+1}} = - \dfrac{\log (2 \sin 2^x \theta)}{2^x}$.

5. $\sum \log (\cos a - \sin a \tan ax) = \log (\cos ax)$.

6. $\sum \log (\cos a + \sin a \cot ax) = \log (\sin ax)$.

G.   Forms Involving Arc Tangents.

1. $\sum \tan^{-1} \dfrac{1}{1 + x + x^2} = \tan^{-1} x$.

2. $\sum \tan^{-1} \dfrac{(ax - x + a)a^x}{1 + (x^2 + x) a^{2x+1}} = \tan^{-1} xa^x$.

3. $\sum \tan^{-1} \left\{ \dfrac{\theta}{1 + \theta^2 x + \theta^2 x^2} \right\} = \tan^{-1}(x\theta)$.

4. $\sum \tan^{-1} \left\{ \dfrac{\theta}{1 + \theta (h + x\theta) + (h + x\theta)^2} \right\} = \tan^{-1}(h + x\theta)$.

5. $\sum \tan^{-1} \dfrac{\Delta u_x}{1 + u_x u_{x+1}} = \tan^{-1} u_x.$

6. $\sum 2^x \tan^{-1} \left\{ \dfrac{\theta^3}{4 \cdot 2^{3x} + 3\theta^2 2^x} \right\} = 2^x \tan^{-1} \left(\dfrac{\theta}{2^x} \right).$

7. $\sum \tan^{-1} \left\{ \dfrac{Ab - aB}{\begin{pmatrix} A^2 + AB \\ +a^2 + ab \end{pmatrix} + x \begin{pmatrix} 2AB + B^2 \\ +2ab + b^2 \end{pmatrix} + x^2 (B^2 + b^2)} \right\}$

$\qquad = \tan^{-1} \dfrac{a + bx}{A + Bx}.$

H. Miscellaneous Forms.

1. $\sum \Psi(x + 1) = x\,\Psi(x) - x.$

2. $\sum \left(\dfrac{y}{x + y} \right) B(x, y) = -B(x, y),$ where $B(x, y) = \dfrac{\Gamma(x)\,\Gamma(y)}{\Gamma(x + y)}.$

3. $\sum \sinh\,(ax + b) = \dfrac{\cosh\,(ax + b - a/2)}{2 \sinh a/2}.$

4. $\sum \cosh\,(ax + b) = \dfrac{\sinh\,(ax + b - a/2)}{2 \sinh a/2}.$

5. $\sum x^{n-1} = B_n(x)/n,$ where $B_n(x)$ is the nth Bernoulli polynomial.

6. $\sum \left[\Delta \Phi_{2n}(x) \right] = \Phi_{2n}(x) = \Phi_{2n}(1 - x),$ where $\Phi_n(x) = B_n(x)/n!.$

7. $\sum \Phi_n(x) = (x - 1)\,\Phi_n(x) - n\,\Phi_{n+1}(x).$

8. $\sum x^2\,\Gamma(x) = \Gamma(x + 1) = x\,\Gamma(x).$

9. $\sum (x + a)^2\,\Gamma(x + a) = \Gamma(x + a + 1) = (x + a)\,\Gamma(x + a).$

10. $\displaystyle\sum \frac{(b+x)!}{(a+x)!} = \frac{1}{(b-a+1)} \frac{(b+x)!}{(a+x-1)!}.$

11. $\displaystyle\sum \frac{(b-x)!}{(a-x)!} = \frac{1}{(a-b-1)} \frac{(b-x+1)!}{(a-x)!}.$

12. $\displaystyle\sum \frac{\Gamma(B+x)}{\Gamma(A+x)} = \frac{1}{(B-A+1)} \frac{\Gamma(B+x)}{\Gamma(A+x-1)}.$

13. $\displaystyle\sum \frac{\Gamma(B-x)}{\Gamma(A-x)} = \frac{1}{(A-B-1)} \frac{\Gamma(B-x+1)}{\Gamma(A-x)}.$

14. $\displaystyle\sum (-1)^x m^{(x)} n^{(-x)} = \frac{(-1)^{x-1}}{m+n} m^{(x)} n^{(-x+1)},$

$$\text{where } m^{(x)} = m(m-1)\cdots(m-x+1),$$

$$n^{(-n)} = 1/[(n+1)(n+2)\cdots(n+x)],$$

and m and n are integers.

INDEX OF NAMES

INDEX OF SUBJECTS

A CATALOG OF SELECTED
DOVER BOOKS
IN SCIENCE AND MATHEMATICS

Mathematics–Bestsellers

HANDBOOK OF MATHEMATICAL FUNCTIONS: with Formulas, Graphs, and Mathematical Tables, Edited by Milton Abramowitz and Irene A. Stegun. A classic resource for working with special functions, standard trig, and exponential logarithmic definitions and extensions, it features 29 sets of tables, some to as high as 20 places. 1046pp. 8 x 10 1/2. 0-486-61272-4

ABSTRACT AND CONCRETE CATEGORIES: The Joy of Cats, Jiri Adamek, Horst Herrlich, and George E. Strecker. This up-to-date introductory treatment employs category theory to explore the theory of structures. Its unique approach stresses concrete categories and presents a systematic view of factorization structures. Numerous examples. 1990 edition, updated 2004. 528pp. 6 1/8 x 9 1/4. 0-486-46934-4

MATHEMATICS: Its Content, Methods and Meaning, A. D. Aleksandrov, A. N. Kolmogorov, and M. A. Lavrent'ev. Major survey offers comprehensive, coherent discussions of analytic geometry, algebra, differential equations, calculus of variations, functions of a complex variable, prime numbers, linear and non-Euclidean geometry, topology, functional analysis, more. 1963 edition. 1120pp. 5 3/8 x 8 1/2. 0-486-40916-3

INTRODUCTION TO VECTORS AND TENSORS: Second Edition--Two Volumes Bound as One, Ray M. Bowen and C.-C. Wang. Convenient single-volume compilation of two texts offers both introduction and in-depth survey. Geared toward engineering and science students rather than mathematicians, it focuses on physics and engineering applications. 1976 edition. 560pp. 6 1/2 x 9 1/4. 0-486-46914-X

AN INTRODUCTION TO ORTHOGONAL POLYNOMIALS, Theodore S. Chihara. Concise introduction covers general elementary theory, including the representation theorem and distribution functions, continued fractions and chain sequences, the recurrence formula, special functions, and some specific systems. 1978 edition. 272pp. 5 3/8 x 8 1/2. 0-486-47929-3

ADVANCED MATHEMATICS FOR ENGINEERS AND SCIENTISTS, Paul DuChateau. This primary text and supplemental reference focuses on linear algebra, calculus, and ordinary differential equations. Additional topics include partial differential equations and approximation methods. Includes solved problems. 1992 edition. 400pp. 7 1/2 x 9 1/4. 0-486-47930-7

PARTIAL DIFFERENTIAL EQUATIONS FOR SCIENTISTS AND ENGINEERS, Stanley J. Farlow. Practical text shows how to formulate and solve partial differential equations. Coverage of diffusion-type problems, hyperbolic-type problems, elliptic-type problems, numerical and approximate methods. Solution guide available upon request. 1982 edition. 414pp. 6 1/8 x 9 1/4. 0-486-67620-X

VARIATIONAL PRINCIPLES AND FREE-BOUNDARY PROBLEMS, Avner Friedman. Advanced graduate-level text examines variational methods in partial differential equations and illustrates their applications to free-boundary problems. Features detailed statements of standard theory of elliptic and parabolic operators. 1982 edition. 720pp. 6 1/8 x 9 1/4. 0-486-47853-X

LINEAR ANALYSIS AND REPRESENTATION THEORY, Steven A. Gaal. Unified treatment covers topics from the theory of operators and operator algebras on Hilbert spaces; integration and representation theory for topological groups; and the theory of Lie algebras, Lie groups, and transform groups. 1973 edition. 704pp. 6 1/8 x 9 1/4. 0-486-47851-3

Browse over 9,000 books at www.doverpublications.com

A SURVEY OF INDUSTRIAL MATHEMATICS, Charles R. MacCluer. Students learn how to solve problems they'll encounter in their professional lives with this concise single-volume treatment. It employs MATLAB and other strategies to explore typical industrial problems. 2000 edition. 384pp. 5 3/8 x 8 1/2. 0-486-47702-9

NUMBER SYSTEMS AND THE FOUNDATIONS OF ANALYSIS, Elliott Mendelson. Geared toward undergraduate and beginning graduate students, this study explores natural numbers, integers, rational numbers, real numbers, and complex numbers. Numerous exercises and appendixes supplement the text. 1973 edition. 368pp. 5 3/8 x 8 1/2. 0-486-45792-3

A FIRST LOOK AT NUMERICAL FUNCTIONAL ANALYSIS, W. W. Sawyer. Text by renowned educator shows how problems in numerical analysis lead to concepts of functional analysis. Topics include Banach and Hilbert spaces, contraction mappings, convergence, differentiation and integration, and Euclidean space. 1978 edition. 208pp. 5 3/8 x 8 1/2. 0-486-47882-3

FRACTALS, CHAOS, POWER LAWS: Minutes from an Infinite Paradise, Manfred Schroeder. A fascinating exploration of the connections between chaos theory, physics, biology, and mathematics, this book abounds in award-winning computer graphics, optical illusions, and games that clarify memorable insights into self-similarity. 1992 edition. 448pp. 6 1/8 x 9 1/4. 0-486-47204-3

SET THEORY AND THE CONTINUUM PROBLEM, Raymond M. Smullyan and Melvin Fitting. A lucid, elegant, and complete survey of set theory, this three-part treatment explores axiomatic set theory, the consistency of the continuum hypothesis, and forcing and independence results. 1996 edition. 336pp. 6 x 9. 0-486-47484-4

DYNAMICAL SYSTEMS, Shlomo Sternberg. A pioneer in the field of dynamical systems discusses one-dimensional dynamics, differential equations, random walks, iterated function systems, symbolic dynamics, and Markov chains. Supplementary materials include PowerPoint slides and MATLAB exercises. 2010 edition. 272pp. 6 1/8 x 9 1/4. 0-486-47705-3

ORDINARY DIFFERENTIAL EQUATIONS, Morris Tenenbaum and Harry Pollard. Skillfully organized introductory text examines origin of differential equations, then defines basic terms and outlines general solution of a differential equation. Explores integrating factors; dilution and accretion problems; Laplace Transforms; Newton's Interpolation Formulas, more. 818pp. 5 3/8 x 8 1/2. 0-486-64940-7

MATROID THEORY, D. J. A. Welsh. Text by a noted expert describes standard examples and investigation results, using elementary proofs to develop basic matroid properties before advancing to a more sophisticated treatment. Includes numerous exercises. 1976 edition. 448pp. 5 3/8 x 8 1/2. 0-486-47439-9

THE CONCEPT OF A RIEMANN SURFACE, Hermann Weyl. This classic on the general history of functions combines function theory and geometry, forming the basis of the modern approach to analysis, geometry, and topology. 1955 edition. 208pp. 5 3/8 x 8 1/2. 0-486-47004-0

THE LAPLACE TRANSFORM, David Vernon Widder. This volume focuses on the Laplace and Stieltjes transforms, offering a highly theoretical treatment. Topics include fundamental formulas, the moment problem, monotonic functions, and Tauberian theorems. 1941 edition. 416pp. 5 3/8 x 8 1/2. 0-486-47755-X

Browse over 9,000 books at www.doverpublications.com

Mathematics–Logic and Problem Solving

PERPLEXING PUZZLES AND TANTALIZING TEASERS, Martin Gardner. Ninety-three riddles, mazes, illusions, tricky questions, word and picture puzzles, and other challenges offer hours of entertainment for youngsters. Filled with rib-tickling drawings. Solutions. 224pp. 5 3/8 x 8 1/2.   0-486-25637-5

MY BEST MATHEMATICAL AND LOGIC PUZZLES, Martin Gardner. The noted expert selects 70 of his favorite "short" puzzles. Includes The Returning Explorer, The Mutilated Chessboard, Scrambled Box Tops, and dozens more. Complete solutions included. 96pp. 5 3/8 x 8 1/2.   0-486-28152-3

THE LADY OR THE TIGER?: and Other Logic Puzzles, Raymond M. Smullyan. Created by a renowned puzzle master, these whimsically themed challenges involve paradoxes about probability, time, and change; metapuzzles; and self-referentiality. Nineteen chapters advance in difficulty from relatively simple to highly complex. 1982 edition. 240pp. 5 3/8 x 8 1/2.   0-486-47027-X

SATAN, CANTOR AND INFINITY: Mind-Boggling Puzzles, Raymond M. Smullyan. A renowned mathematician tells stories of knights and knaves in an entertaining look at the logical precepts behind infinity, probability, time, and change. Requires a strong background in mathematics. Complete solutions. 288pp. 5 3/8 x 8 1/2.

0-486-47036-9

THE RED BOOK OF MATHEMATICAL PROBLEMS, Kenneth S. Williams and Kenneth Hardy. Handy compilation of 100 practice problems, hints and solutions indispensable for students preparing for the William Lowell Putnam and other mathematical competitions. Preface to the First Edition. Sources. 1988 edition. 192pp. 5 3/8 x 8 1/2.   0-486-69415-1

KING ARTHUR IN SEARCH OF HIS DOG AND OTHER CURIOUS PUZZLES, Raymond M. Smullyan. This fanciful, original collection for readers of all ages features arithmetic puzzles, logic problems related to crime detection, and logic and arithmetic puzzles involving King Arthur and his Dogs of the Round Table. 160pp. 5 3/8 x 8 1/2.   0-486-47435-6

UNDECIDABLE THEORIES: Studies in Logic and the Foundation of Mathematics, Alfred Tarski in collaboration with Andrzej Mostowski and Raphael M. Robinson. This well-known book by the famed logician consists of three treatises: "A General Method in Proofs of Undecidability," "Undecidability and Essential Undecidability in Mathematics," and "Undecidability of the Elementary Theory of Groups." 1953 edition. 112pp. 5 3/8 x 8 1/2.   0-486-47703-7

LOGIC FOR MATHEMATICIANS, J. Barkley Rosser. Examination of essential topics and theorems assumes no background in logic. "Undoubtedly a major addition to the literature of mathematical logic." – *Bulletin of the American Mathematical Society.* 1978 edition. 592pp. 6 1/8 x 9 1/4.   0-486-46898-4

INTRODUCTION TO PROOF IN ABSTRACT MATHEMATICS, Andrew Wohlgemuth. This undergraduate text teaches students what constitutes an acceptable proof, and it develops their ability to do proofs of routine problems as well as those requiring creative insights. 1990 edition. 384pp. 6 1/2 x 9 1/4.   0-486-47854-8

FIRST COURSE IN MATHEMATICAL LOGIC, Patrick Suppes and Shirley Hill. Rigorous introduction is simple enough in presentation and context for wide range of students. Symbolizing sentences; logical inference; truth and validity; truth tables; terms, predicates, universal quantifiers; universal specification and laws of identity; more. 288pp. 5 3/8 x 8 1/2.   0-486-42259-3

Browse over 9,000 books at www.doverpublications.com

Mathematics–Algebra and Calculus

VECTOR CALCULUS, Peter Baxandall and Hans Liebeck. This introductory text offers a rigorous, comprehensive treatment. Classical theorems of vector calculus are amply illustrated with figures, worked examples, physical applications, and exercises with hints and answers. 1986 edition. 560pp. 5 3/8 x 8 1/2. 0-486-46620-5

ADVANCED CALCULUS: An Introduction to Classical Analysis, Louis Brand. A course in analysis that focuses on the functions of a real variable, this text introduces the basic concepts in their simplest setting and illustrates its teachings with numerous examples, theorems, and proofs. 1955 edition. 592pp. 5 3/8 x 8 1/2. 0-486-44548-8

ADVANCED CALCULUS, Avner Friedman. Intended for students who have already completed a one-year course in elementary calculus, this two-part treatment advances from functions of one variable to those of several variables. Solutions. 1971 edition. 432pp. 5 3/8 x 8 1/2. 0-486-45795-8

METHODS OF MATHEMATICS APPLIED TO CALCULUS, PROBABILITY, AND STATISTICS, Richard W. Hamming. This 4-part treatment begins with algebra and analytic geometry and proceeds to an exploration of the calculus of algebraic functions and transcendental functions and applications. 1985 edition. Includes 310 figures and 18 tables. 880pp. 6 1/2 x 9 1/4. 0-486-43945-3

BASIC ALGEBRA I: Second Edition, Nathan Jacobson. A classic text and standard reference for a generation, this volume covers all undergraduate algebra topics, including groups, rings, modules, Galois theory, polynomials, linear algebra, and associative algebra. 1985 edition. 528pp. 6 1/8 x 9 1/4. 0-486-47189-6

BASIC ALGEBRA II: Second Edition, Nathan Jacobson. This classic text and standard reference comprises all subjects of a first-year graduate-level course, including in-depth coverage of groups and polynomials and extensive use of categories and functors. 1989 edition. 704pp. 6 1/8 x 9 1/4. 0-486-47187-X

CALCULUS: An Intuitive and Physical Approach (Second Edition), Morris Kline. Application-oriented introduction relates the subject as closely as possible to science with explorations of the derivative; differentiation and integration of the powers of x; theorems on differentiation, antidifferentiation; the chain rule; trigonometric functions; more. Examples. 1967 edition. 960pp. 6 1/2 x 9 1/4. 0-486-40453-6

ABSTRACT ALGEBRA AND SOLUTION BY RADICALS, John E. Maxfield and Margaret W. Maxfield. Accessible advanced undergraduate-level text starts with groups, rings, fields, and polynomials and advances to Galois theory, radicals and roots of unity, and solution by radicals. Numerous examples, illustrations, exercises, appendixes. 1971 edition. 224pp. 6 1/8 x 9 1/4. 0-486-47723-1

AN INTRODUCTION TO THE THEORY OF LINEAR SPACES, Georgi E. Shilov. Translated by Richard A. Silverman. Introductory treatment offers a clear exposition of algebra, geometry, and analysis as parts of an integrated whole rather than separate subjects. Numerous examples illustrate many different fields, and problems include hints or answers. 1961 edition. 320pp. 5 3/8 x 8 1/2. 0-486-63070-6

LINEAR ALGEBRA, Georgi E. Shilov. Covers determinants, linear spaces, systems of linear equations, linear functions of a vector argument, coordinate transformations, the canonical form of the matrix of a linear operator, bilinear and quadratic forms, and more. 387pp. 5 3/8 x 8 1/2. 0-486-63518-X

Browse over 9,000 books at www.doverpublications.com

Mathematics–Probability and Statistics

BASIC PROBABILITY THEORY, Robert B. Ash. This text emphasizes the probabilistic way of thinking, rather than measure-theoretic concepts. Geared toward advanced undergraduates and graduate students, it features solutions to some of the problems. 1970 edition. 352pp. 5 3/8 x 8 1/2. 0-486-46628-0

PRINCIPLES OF STATISTICS, M. G. Bulmer. Concise description of classical statistics, from basic dice probabilities to modern regression analysis. Equal stress on theory and applications. Moderate difficulty; only basic calculus required. Includes problems with answers. 252pp. 5 5/8 x 8 1/4. 0-486-63760-3

OUTLINE OF BASIC STATISTICS: Dictionary and Formulas, John E. Freund and Frank J. Williams. Handy guide includes a 70-page outline of essential statistical formulas covering grouped and ungrouped data, finite populations, probability, and more, plus over 1,000 clear, concise definitions of statistical terms. 1966 edition. 208pp. 5 3/8 x 8 1/2. 0-486-47769-X

GOOD THINKING: The Foundations of Probability and Its Applications, Irving J. Good. This in-depth treatment of probability theory by a famous British statistician explores Keynesian principles and surveys such topics as Bayesian rationality, corroboration, hypothesis testing, and mathematical tools for induction and simplicity. 1983 edition. 352pp. 5 3/8 x 8 1/2. 0-486-47438-0

INTRODUCTION TO PROBABILITY THEORY WITH CONTEMPORARY APPLICATIONS, Lester L. Helms. Extensive discussions and clear examples, written in plain language, expose students to the rules and methods of probability. Exercises foster problem-solving skills, and all problems feature step-by-step solutions. 1997 edition. 368pp. 6 1/2 x 9 1/4. 0-486-47418-6

CHANCE, LUCK, AND STATISTICS, Horace C. Levinson. In simple, non-technical language, this volume explores the fundamentals governing chance and applies them to sports, government, and business. "Clear and lively ... remarkably accurate." – *Scientific Monthly*. 384pp. 5 3/8 x 8 1/2. 0-486-41997-5

FIFTY CHALLENGING PROBLEMS IN PROBABILITY WITH SOLUTIONS, Frederick Mosteller. Remarkable puzzlers, graded in difficulty, illustrate elementary and advanced aspects of probability. These problems were selected for originality, general interest, or because they demonstrate valuable techniques. Also includes detailed solutions. 88pp. 5 3/8 x 8 1/2. 0-486-65355-2

EXPERIMENTAL STATISTICS, Mary Gibbons Natrella. A handbook for those seeking engineering information and quantitative data for designing, developing, constructing, and testing equipment. Covers the planning of experiments, the analyzing of extreme-value data; and more. 1966 edition. Index. Includes 52 figures and 76 tables. 560pp. 8 3/8 x 11. 0-486-43937-2

STOCHASTIC MODELING: Analysis and Simulation, Barry L. Nelson. Coherent introduction to techniques also offers a guide to the mathematical, numerical, and simulation tools of systems analysis. Includes formulation of models, analysis, and interpretation of results. 1995 edition. 336pp. 6 1/8 x 9 1/4. 0-486-47770-3

INTRODUCTION TO BIOSTATISTICS: Second Edition, Robert R. Sokal and F. James Rohlf. Suitable for undergraduates with a minimal background in mathematics, this introduction ranges from descriptive statistics to fundamental distributions and the testing of hypotheses. Includes numerous worked-out problems and examples. 1987 edition. 384pp. 6 1/8 x 9 1/4. 0-486-46961-1

Browse over 9,000 books at www.doverpublications.com